FUNDAMENTALS OF LASER MICROMACHINING

FUNDAMENTALS OF LASER MICROMACHINING

Ronald D. Schaeffer

CRC Press is an imprint of the
Taylor & Francis Group, an **informa** business
A TAYLOR & FRANCIS BOOK

Cover image courtesy of CCAT (Connecticut Center for Advanced Technology)

CRC Press
Taylor & Francis Group
6000 Broken Sound Parkway NW, Suite 300
Boca Raton, FL 33487-2742

Printed in the United States of America on acid-free paper
Version Date: 20120316

International Standard Book Number: 978-1-4398-6055-7 (Hardback)

Library of Congress Cataloging-in-Publication Data

Schaeffer, Ronald D.
Fundamentals of laser micromachining / Ronald D. Schaeffer.
p. cm.
"A CRC title."
Includes bibliographical references and index.
ISBN 978-1-4398-6055-7 (alk. paper)
1. Lasers--Industrial applications. I. Title.

TA1675.S33 2012
671.3'5--dc23 2011045634

Visit the Taylor & Francis Web site at
http://www.taylorandfrancis.com

and the CRC Press Web site at
http://www.crcpress.com

Contents

List of Figures

Acknowledgments

I would like to extend my thanks to the many friends and colleagues who have helped me in many ways to make this book possible. First, a note of thanks to the publisher and, specifically, John Navas, for bugging me for several years to make this happen while I dragged my feet. I hope that the additional material developed in the interim makes the final product better than it otherwise would have been.

I have tried to avoid using particular company names or product names wherever possible to keep the information presented as generic as possible. Of course, I do have some personal prejudices in this regard and it is probably obvious in many cases to which specific products or companies I am referring. This is difficult to avoid completely. Figures that have been provided by entities other than PhotoMachining, Inc. have the acknowledgments noted in the figure captions.

All of my colleagues at PhotoMachining, Inc. have contributed to this work in some way or another. They deserve the credit for most of the figures and the actual work was performed by them. In particular, Leah Guerrero, Gabor Kardos, Dr. Oleg Derkach, David Grossman, and Scott Crandall provided many of the photos taken over the years for this book and my many other publications.

In addition, text reviews and valuable advice were given by John O'Connell, Dr. Francis (Bo) Burns, Sidney Wright, Dr. Dirk Muller, Andreas Engelmayer, Dr. Tony Hoult, Jeremy Donatell, Todd Lizotte, Dr. Ken Dzurko, Jenifer Bunis, Leonard Migliore, David Belforte, and Gus Anibarro. Figures were graciously provided by Clark MXR, Coherent, Raydiance, JPSA, Synrad, Hitachi Via Mechanics USA, Lizotte Tactical LLC, CCAT (Connecticut Center for Advanced Technology), Lumera, LPL Systems, and LPKF.

Dagmar Schaefer, Dr. Ingomar Kelbassa, and the Fraunhofer ILT team provided figures and information on ISLE, and some of this work was published under joint authorship in *MICROmanufacturing* magazine. Other portions of the text were also published in various articles and columns written over the years (some with coauthors) for a variety of publications, mostly *MICROmanufacturing* and *Industrial Laser Solutions.* The work on multiple hole drilling using galvos was part of an SBIR phase I research effort supported by the US Air Force.

Finally, I would like to thank Meghan Szmyt and Martin Gastrock for their invaluable help in formatting the text, arranging the figures, and doing all of the things necessary to make my thoughts cohesive and presentable.

The Author

Ronald Schaeffer is chief executive officer of PhotoMachining, Inc. He has been involved in laser manufacture and materials processing for over 25 years, working in and starting small companies. He has over 130 publications, has written monthly web and print columns (currently writing a column for *MICROmanufacturing* magazine), and is on the editorial advisory board of *Industrial Laser Solutions*. He is also a past member of the board of directors of the Laser Institute of America and is affiliated with the New England Board of Higher Education. He has a PhD in physical chemistry from Lehigh University and did graduate work at the University of Paris, after which he worked for several major laser companies. He is a US Army veteran of the 172nd Mountain Brigade and the 101st Airborne Division. In his spare time, he farms, collects antique pocket watches, plays guitar, and rides motorcycles.

1

Introduction

Lasers were first demonstrated in the 1950s in the microwave region of the electromagnetic spectrum, and this subset of lasers was called a MASER (Microwave Amplified by Stimulated Emission of Radiation). By 1964 most of the lasers that we know today had already been conceived, but it took many more years for lasers to be accepted into industrial environments. One reason is that early lasers were highly unreliable, but possibly just as important was the fact that lasers were solutions to problems that nobody, at that time, knew existed.

Today, lasers are ubiquitous inside our homes and also on the manufacturing floor. Lasers are used in a wide variety of applications: to cut and weld metal for building ships, automobiles, and railroad cars; to perform surgery; to put on fabulous light shows; to gather forensic evidence; and to measure the distance from earth to the moon accurately. Lasers are also used to make very small features in materials. These small features are critical to the electronics, aerospace, medical, renewable energy, automobile, and other industries.

The machining of these small features is called *laser micromachining*—the subject of this book. The definition of laser micromachining used in this text is the following: the machining is done using lasers (instead of, for instance, a drill bit or saw blade), there is material removal or at least modification (not including welding or additive processes, which is another large market for lasers), the feature sizes are less than 1 mm (usually a lot less), and the thickness of material is also less than 1 mm (usually a lot less). Typically, the smallest attainable feature sizes are on the order of a micron. There are some exceptions to these rules, but not many.

The field of lasers encompasses a broad range of different topics, which are beyond the scope of this book. Enough theoretical background will be presented so that the reader can understand the basic concepts. If the reader is interested in more detailed theoretical discussions, these can be found in the references cited at the end of this text.

What is a laser? A laser is a device that generates or amplifies light. The acronym LASER stands for Light Amplified by Stimulated Emission of Radiation. The term was coined by Dr. Gordon Gould in his laboratory notebook when he was a student of Dr. Charles Townes (Nobel Prize winner and co-inventor of the MASER) at Columbia University in 1957.

There are three fundamental, essential elements of a LASER: the lasing medium (gas, liquid, or solid), the pumping process, and optical feedback elements. Industrial lasers are for the most part either gas or solid state as the liquids used in lasers tend to be hazardous, difficult to work with, and not typically suited for the workplace. Therefore, this text confines discussions to solid- and gas-phase lasers. The pumping process simply puts energy into the system. This input energy is used to create a *population inversion:* a state where the upper state transitional energy level has more resident species than the lower state transitional energy level.

Population inversion is critical to sustaining laser operation and is in principle impossible to achieve by pumping directly from the lower laser transition level to the upper laser transition level. Therefore, most lasing quantum transitions involve multilevel populating and depopulating schemes. If the laser cavity is designed correctly, the resulting light is monochromatic, directional, intense, and coherent. While coherence is very important in some applications, such as interferometry, it has not been found to be important in machining, so it is not addressed in detail in this text.

Monochromatic means single wavelength. Lasers output either a single wavelength (with some associated line width) or several wavelengths that can be filtered to give a single wavelength output. This single wavelength characteristic is important because materials interact in different ways to light of different wavelengths (colors) and, in many cases, this allows one to perform selective material removal by using the correct wavelength of light. Single wavelength output also makes the optical setup easier and allows a tight spot to be focused on target. *Directional* means that the output beam moves in one direction and the beam spreads very little over distance—a low divergence beam. *Intensity* or brightness defines the high density of usable photons available from lasers.

Why use lasers for material removal? First, lasers are noncontact. Therefore, there is less chance of mechanical damage to the parts being processed. Second, and perhaps more important, there is no tool wear as seen in traditional processes like EDM (Electrode Discharge Machining) or milling/drilling machines. Lasers can also be a one-step process, unlike, for instance, etching, which involves toxic chemicals and is a multistep process. By varying the wavelength and energy density on target, one can also get selective material removal—for example, in removing dielectrics from conductors. Finally, lasers provide a very flexible tool since advanced systems incorporate computer control with programming interfaces that permit "soft tooling"; this is especially valuable for prototyping where high tool-up costs must be avoided.

Table 1.1 shows the capabilities of some lasers compared to other "traditional" machining methods. All of the lasers can make features with sizes less than 100 μm and some can produce much smaller features. Other techniques are pretty much at their usable limit at 100 μm feature size. This is not to say that some of these techniques cannot be used for smaller features, but rather that the cost to do so goes up significantly. Mechanical drilling or milling is pretty straightforward and involves using a hard tool to cut

TABLE 1.1

Comparison of Machining Methods

	Practical Resolution Limit	Attainable Aspect Ratio[a]	Taper	Undesirable Side Effects	Status of Technology Development
Excimer	2 μm	>100:1	Yes	Recast layer	Moderate
CO_2 laser	75 μm	100:1	Yes	Recast layer, burring, thermal	High
Nd:YAG	10 μm	100:1	Yes	Recast layer, burring, thermal	High
EDM	100 μm	20:1	No	Surface finish	Moderate
Chemical etch	200 μm	1:1.5	Yes	Undercutting	Moderate
Mechanical	100 μm	10:1	No	Burring	High

[a] Depth/hole diameter.

features. This is a serial process when using a single bit machine and like most lasers in that one hole is made at a time, so speed is a consideration.

On the other hand, chemical etching is a batch process where *all* of the features are made simultaneously. This process, though, requires the use of toxic chemicals, and also the attainable aspect ratio (depth/hole diameter) is on the order of 1:1, so if a 50 μm diameter hole in a 200 μm thick piece of material is required, an etch process cannot be used. Electric discharge machining (EDM) is a fine way to process metals and, since they can be stacked, high processing speeds can be achieved. The drawback is that the material must be conductive in order for EDM to work.

By using lasers, it is possible to complement and in some way supplement the more traditional technologies mentioned before. Laser micromachining is instrumental in manufacturing a wide variety of products, from medical implants to aircraft engines, from printed circuit boards to gyroscopes, from fuel cells to drug delivery catheters.

However, a "LASER" is not a "LASER"! Many different types of lasers are commercially available and they can be used in a variety of different ways. In order to understand where a particular laser is best utilized, it is helpful to review some background material on laser physics and optics as well as the hardware necessary to take the laser (which by itself is nothing more than a glorified light bulb) and make it into a production tool. Once this basic understanding is in place, a better understanding of the impact on markets and materials can be achieved.

Finally, this text is not intended to be exhaustive. The author surely has some prejudices and it is certain they will be noted throughout, so inclusion of some topics or exclusion of other topics may be a matter of personal experience rather than an indication of the relative importance of any particular topic. For instance, the author has extensive experience and background in ultraviolet wavelength lasers more than any other wavelength, so hopefully latitude will be given in this respect. Also, this body of work has been supported by many colleagues over 25 years in the laser micromachining industry and the acknowledgments given at the beginning of the book are surely not comprehensive.

Problems intended to further understanding appear at the end of the book.

2

Laser Theory and Operation

2.1 Brief Review of Laser Physics

In order to facilitate understanding of the micromachining process, it is useful to discuss some details of the theory of laser operation and the definition of some terms used and also to look more closely at the available lasers. This treatment is not intended to be in depth and more information is presented in the Appendix.

2.1.1 Quantum Theory of Light

The quantum theory of light was developed by Planck and Einstein in the early 1900s. The theory states that light is quantized in discrete bundles of energy called "photons." These photons are emitted when atoms or molecules drop from an excited energy state to a lower state. Each light photon has an associated energy that depends on its frequency:

$$E_{\text{photon}} = \text{h}\nu = E_2 - E_1 \tag{2.1}$$

where E_2 and E_1 are upper and lower state energy levels, respectively, h is Planck's constant ($h = 6.626 \times 10^{-34}$ J–s), and ν is the frequency of oscillation (s^{-1}).

Although light is packaged in discrete photons (particle theory of light), light also is characterized by frequency and wavelength λ (wave theory of light):

$$\lambda = c/\nu \tag{2.2}$$

where c = speed of light = 3×10^8 m/s. Consequently, photon energy is proportional to frequency but inversely proportional to wavelength.

2.1.2 Photon Interactions with Matter

Photons can interact with matter in many ways. The most important for micromachining applications include the following:

- *Transmission.* All or some of the photons go through a material, perhaps associated with decrease in energy or change of direction. Transmissive materials make good windows.
- *Absorption.* The photon is absorbed by an atom or molecule, resulting in the photon energy being used either to increase the total energy of the system (in other words, it heats up) or to eject material. If laser machining is to take place, there must be absorption, and the higher the absorption the better the process efficiency. Absorption is described by Beer's law, $\varepsilon = abc$ (absorption = absorptivity × path length × concentration).
- *Scattering.* The photon is scattered either elastically (without energy loss) or inelastically (with energy loss). *Reflection* is a type of scattering.

2.1.3 Laser Physics and Population Inversion

Light can interact with matter in two more ways that are directly related to the lasing process:

- *Spontaneous emission.* The atom or molecule spontaneously drops to a lower energy state, giving off a photon. This is common for species that have in some way been energetically excited as all excited species prefer to return to their lowest available energy state.
- *Stimulated emission.* The photon stimulates an excited state atom, causing it to emit an additional photon with identical characteristics to the stimulating photon; this is the basis for lasers.

Figure 2.1 diagrams some of these photon interactions.

Energy is stored in molecules according to their electronic configuration and degrees of freedom associated with their physical construction. The degrees of freedom associated with simple molecules in order of increasing energy storage are listed in the following list:

- Translational motion = 3 degrees
- Rotational motion = 3 degrees for all molecules except linear molecules (2 for linear molecules)—stimulated by microwave photons
- Vibrational motion $3N - 6$ degrees ($3N - 5$ for linear molecules), where N = the number of atoms in a molecule—stimulated by infrared (IR) photons
- Outer electron excitation < 20 eV—stimulated by ultraviolet (UV) photons
- Inner electron excitation > 20 eV
- Nuclear excitation ~ 1 MeV

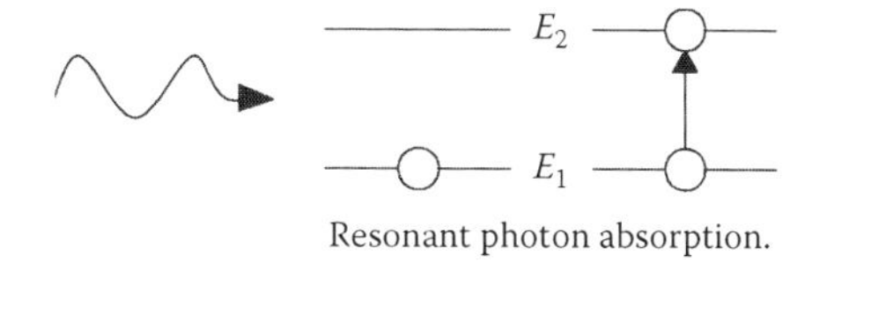

Resonant photon absorption.

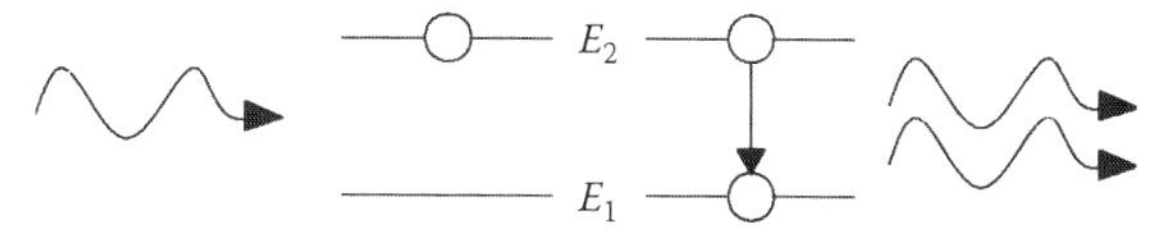

Stimulated photon emission. Energy, wavelength, phase, and direction of a stimulated photon are identical to the incident photon.

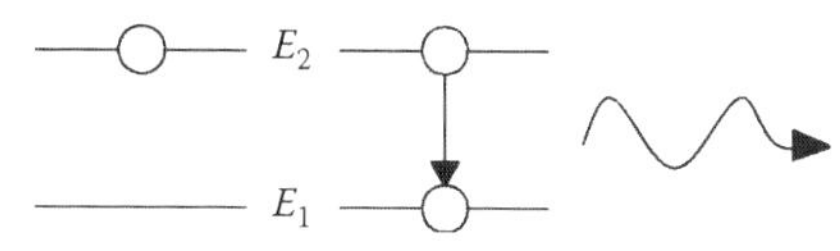

Spontaneous photon emission.

FIGURE 2.1
Photon absorption and stimulated emission.

For instance, CO_2 is a linear molecule containing three atoms and therefore has three translational, two rotational, and four vibrational modes.

A laser requires a "population inversion" to sustain output. A population inversion exists between two lasing energy states when there are more species (N) occupying the upper state than the lower. The process used to excite lower energy atoms or molecules to their excited states is called "pumping." In principle, lasing only requires two energy levels (assuming that the species is created in the excited state), but in practice it is very difficult to populate an upper state energy level with more species than in the true ground state. Thus, in most cases, a four-level pumping scheme is used because it has been shown experimentally that four-level pumping schemes are easier to use than three-level schemes. Requirements for a population inversion in a four-level system include

- Short lifetime τ_3 for state E_3
- Short lifetime τ_1 for state E_1
- High probability of stimulated emission in the laser medium (i.e., high stimulated emission cross section $\sigma_{\text{stim emission}}$)

A population inversion exists when $N_2 > N_1$, as in Figure 2.2. Consequently, the laser pumping mechanism must be sufficient to replace those excited atoms or molecules undergoing spontaneous emission from E_2 and those

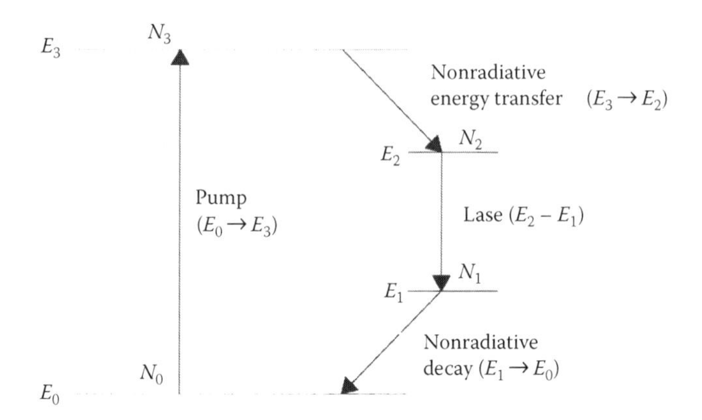

FIGURE 2.2
Population inversion in a four-level system ($N_2 > N_1$).

photons encountering absorption from state E_1 to E_2. Furthermore, if nonradiative decay lifetimes τ_1 or τ_3 are too long, an insufficient population in the upper energy state will result and will not support lasing action.

2.1.4 Essential Elements of a Laser Oscillator

A laser requires a lasing medium, pump process, and a resonator cavity to sustain lasing action, as shown in Figure 2.3. The lasing medium can be a gas, solid, liquid, or semiconductor. The pump process excites the atoms or molecules of the lasing medium to an upper energy state. Lasing action is initiated by spontaneous emission and amplified by stimulated emission along the axis of the resonator cavity. The cavity mirrors reflect the photons back and forth through the laser medium for increased amplification.

Two primary optics—front and rear resonator mirrors—define the laser cavity. Normally, the rear resonator mirror is 100% reflective. The front optic is partially reflective and partially transmissive. Depending on the laser, the front optic reflectivity varies; for a helium-neon laser, the front optic is about 98.5% reflective, but for an excimer laser, the front optic is only 10% reflective. Resonator optics can be concave or otherwise curved,

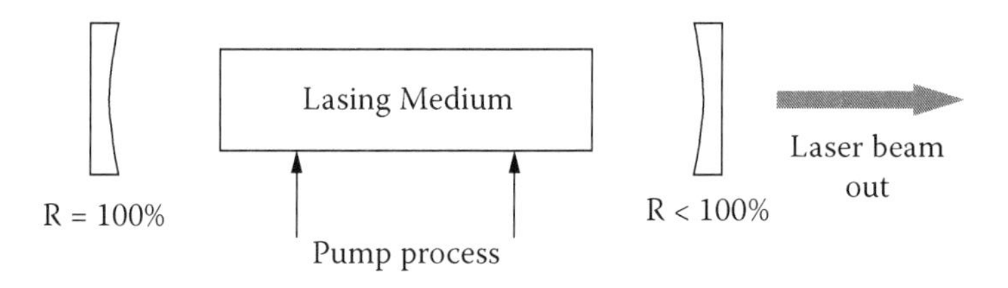

FIGURE 2.3
Essential elements of a laser.

as in a helium-neon laser, or flat, as in an excimer laser. The lasing medium must have a high stimulated emission cross section so that more photons are produced than absorbed. Methods of laser pumping include gas discharge, optical (flashlamp or laser diode), chemical pump, x-rays, and electron beam excitation.

Energy is introduced into the laser through the pumping process, but only a fraction of the "wall plug" energy (the total power used to drive the laser) is present in the laser beam as it exits the front aperture. A typical laser might be less than 10% efficient and, in general, these efficiencies range from about 2% to 40%. Most of the energy is lost in the form of heat.

Because of the shorter wavelength, UV photons have more photon energy than IR photons and a smaller on-target spot size is achievable. The electromagnetic spectrum shown in Figure 2.4 shows some available laser wavelengths.

Why use UV lasers? First, they have a short wavelength (<400 nm), which means that they can be focused to a small spot size (25 μm or less is easy). Second, UV photons are generally absorbed within fractions of a micron of the surface. This allows very fine control of the material removal process, especially when making "blind" features—features that do not go all the way through a material. Since there is less material removal per pulse, processing rates are slower. UV lasers also generally have a relatively short pulse length (<100 ns) and this helps to achieve a high power density on target. Finally, the high photon energy is enough to initiate bond breaking, especially in polymers containing π-electron clouds (double and triple bonds). The minimal thermal effects result in good edge quality. While in principle UV lasers can be used on almost any material, they excel in the processing of polymers.

Why use IR lasers? First, it is important to distinguish between near-IR (1 μm wavelength) and far-IR (10 μm wavelength) lasers. Even though both are IR lasers, they behave differently and interact with materials differently; some materials absorb better in the far IR and some absorb better in the near IR. These differences will be discussed in detail in later sections, but there are also some similarities. IR lasers have relatively long pulse durations (microseconds to milliseconds) and wavelengths. A consequence of the long wavelength is that absorption occurs far into the material (sometimes hundreds of microns per pulse) and also that these lasers are available in very high output powers—up to tens of kilowatts. Another consequence is that, because IR lasers interact with the rotation–vibration modes of molecules and thereby generate heat, IR laser processing is a first-order thermal process. IR lasers have a much more favorable dollars per watt ($/W) ratio than UV lasers. Therefore, IR lasers are best used when speed, part thickness, and large feature sizes are needed; UV lasers are best used when small spot sizes and good edge quality are more important than cost. Table 2.1 lists some common IR and UV micromachining lasers.

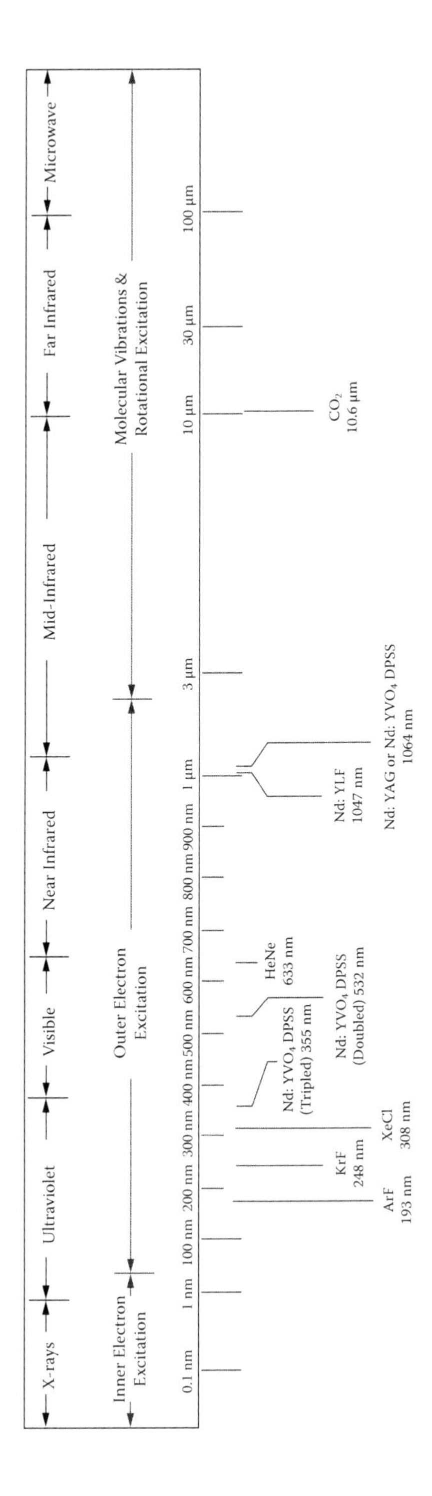

FIGURE 2.4
The electromagnetic spectrum.

TABLE 2.1

Common Lasers Used in the Industry

Type	Medium	Wavelength
Gas lasers	Excimer	193–351 nm
Gas lasers	CO_2	10 μm
Solid-state lasers	Nd:YAG (fundamental)	1.064 μm
	Nd:YAG (second harmonic)	532 nm
	Nd:YAG (third harmonic)	355 nm
	Nd:YAG (fourth harmonic)	266 nm
Fiber lasers	Doped fibers	1064–2000 nm
Disk lasers	Doped thin disks	1064–1030 nm

2.1.5 Important Characteristics of a Laser Beam

Divergence. Any light that exits a confined space will undergo divergence (Figure 2.5). In laser physics, divergence is the degree of spreading that a laser beam exhibits after it exits the front aperture. In machining applications, divergence is undesirable because it leads to reduced energy and distorted images at the target surface, resulting in inconsistent processing results from one part to the next.

Directionality. This property, along with low divergence, is what makes all the photons head in the same direction after exiting the laser oscillator.

Monochromaticity (single wavelength emission). This is somewhat idealized in that most lasers actually emit many wavelengths and a line width is also associated with any laser transition.

Wavelength (or color). The wavelength of a sinusoidal wave is the spatial period of the wave—the distance over which the wave repeats—received from peaks, valleys, or crossover points. Wavelength is inversely proportional to frequency; waves with higher frequency have shorter wavelength and waves with lower frequency have longer wavelength. Figure 2.6 shows a simple wave and Figure 2.7

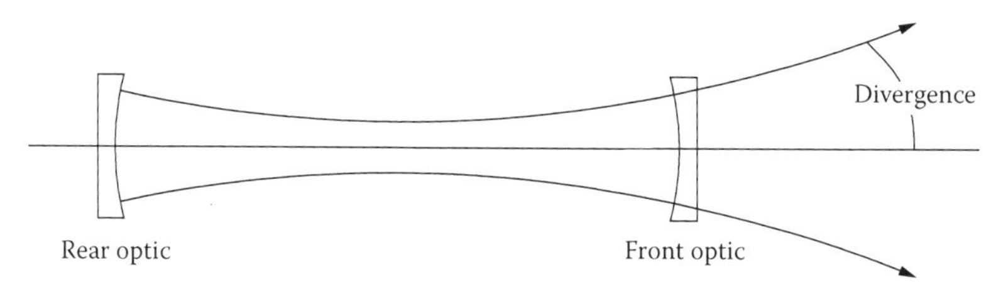

FIGURE 2.5
Divergence characteristics of a laser beam, normally given in units of milliradians.

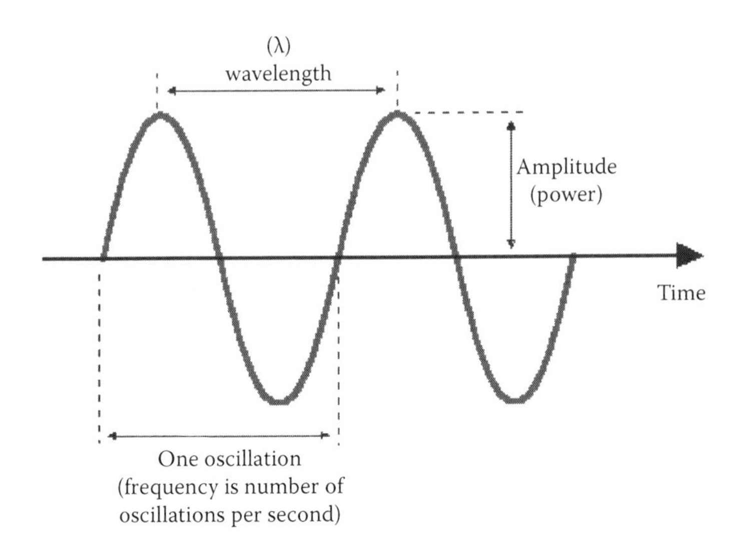

FIGURE 2.6
A simple wave.

depicts the relationship between wavelength and frequency in the electromagnetic spectrum. The part of the electromagnetic spectrum where light output occurs is normally expressed in microns (IR) or nanometers (UV). In general, molecules absorb only specific and distinct wavelengths of light.

Intensity. The density of photons on target, normally expressed in watts per square centimeter (W/cm^2).

Pointing stability. The directional output of a laser beam is subject to fluctuations from mechanical and thermal actions occurring in the resonator and also when propagating through the optical system. This problem becomes worse as the optical path gets longer. With a good laser design, the angular beam-pointing fluctuations should be a tiny fraction of the divergence.

Peak power. This is pulse energy per pulse duration (J/s = W)

Peak power intensity. This is peak power per unit area (spot size) on target (W/cm^2). Peak power intensity seems to be the *key* to clean and low taper processing.

Pulse length (temporal profile). The time duration of the laser pulse that encompasses both the rise and fall times, where "tails" are common. Figure 2.8 shows a typical laser temporal profile. The pulse length is generally described as the width of the curve at half height. Shorter pulses are better for micromachining.

Repetition rate (pulse repetition frequency or prf). This is the number of pulses per second (s^{-1} or Hz) and is the inverse of temporal pulse spacing.

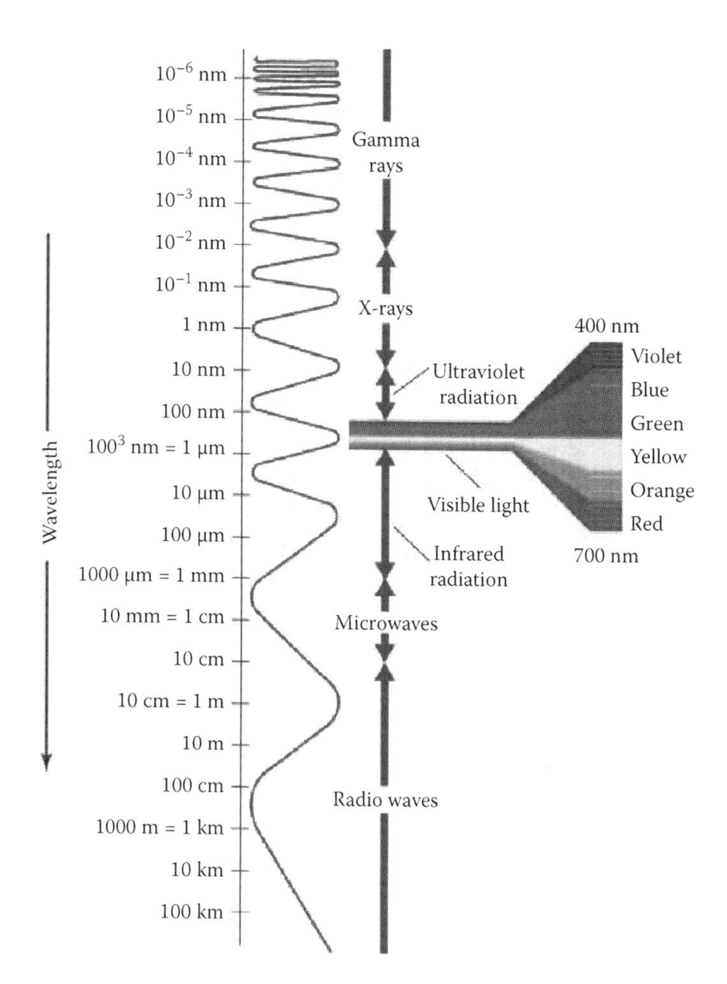

FIGURE 2.7
The relationship between wavelength and frequency in the electromagnetic spectrum.

2.2 CO_2 Lasers

2.2.1 Characteristics of Carbon Dioxide Lasers

CO_2 lasers are far-IR lasers and they are the most common laser in industry. They are relatively inexpensive ($/W) and are available in a wide range of output power, from 10 W to 10 kW. They have a high wall-plug efficiency of over 25%. The output is a series of emission lines from 9.3 to 11.0 µm in the IR with the band center around 10 µm. These emission lines can be filtered by using dielectric optics and/or isotopic mixtures in the gas mix in order to enhance certain lines for better material processing. For instance, 9.3 µm is more efficient for polymer processing and 10.6 µm is better for ceramics.

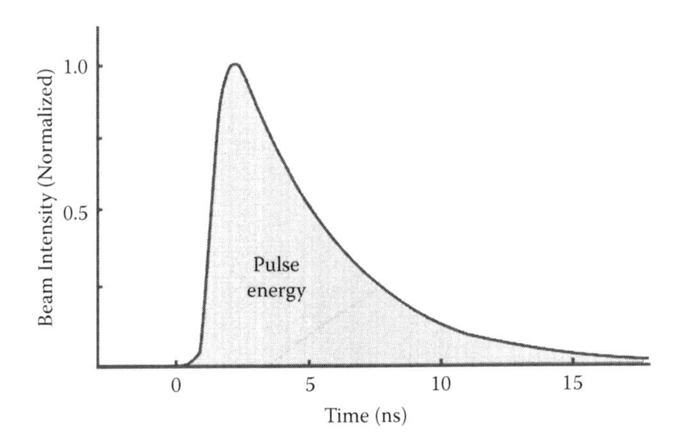

FIGURE 2.8
A typical laser temporal profile.

They have a high penetration depth (5–100 μm or more) and machining is via a first-order thermal process; material/photon interaction is primarily via vibrational excitation. Most of these lasers are used in focal point machining mode except for CO_2–TEA (transverse excited atmospheric) lasers, which have a longitudinal electrode discharge resulting in a high-order multimode beam that is better used in an imaging mode than in focal point machining.

2.2.2 CO_2 Laser Operational Theory

A carbon dioxide laser uses a gas mixture primarily consisting of CO_2:N_2:He, although other constituents are sometimes also found in the gas mixture. The CO_2 molecules constitute the active lasing medium, the N_2 gas serves in an energy transfer mechanism, and the He atoms enhance the population inversion by depopulating the lower energy states. The population inversion and lasing transition in a CO_2 laser is established between vibrational and rotational energy states. Most CO_2 lasers are pumped by a high-pressure electrical discharge.

Excited $CO_2(001)$ molecules are formed by three methods:

- Inelastic collision between electrons and ground-state CO_2 molecules:

$$e^- + CO_2(000) \rightarrow CO_2(001) \tag{2.3}$$

- Vibrational–vibrational excitation via N_2 molecules (laser gas mixture of CO_2:N_2:He)

1. Electrons in the gas discharge current excite N_2 molecules vibrationally by inelastic collision:

$$e^- + N_2(v = 0) \rightarrow N_2(v = n), n = 1 \text{ to } 8 \quad (2.4)$$

2. Excited N_2^* molecules transfer energy to CO_2 molecules:

$$N_2^* (v = n) + CO_2 (000) \rightarrow N_2 + CO_2 (001) \quad (2.5)$$

Electronic excitation and ionization are minor contributors.

Theory and experiment show that 60% of wallplug power can be channeled into pumping the upper CO_2 laser level, resulting in up to 27% wall plug efficiency. Intervibrational energy transfers from N_2 account for this efficiency.

2.2.3 Types of CO_2 Lasers

Most CO_2 lasers have a Gaussian beam. This is essentially a circular beam energy profile. If one takes a linear profile along the diameter, the typical Gaussian profile (otherwise known as a "bell-shaped curve") results, with the hottest spot in the middle and energy tailing off in the wings. The most important CO_2 machining characteristic is that the material interaction occurs via thermal overload and vaporization. Penetration depths are 5–10 μm or more—sometimes much more, depending on the material, optical setup, and the laser power. The diffraction limit is about 10 μm. However, considering optics and processing conditions, the best practical feature resolution is about 50 μm and most lasers and optical setups result in a spot size of greater than 100 μm. Gaussian lasers are used in a focal point machining configuration and can therefore also use a coaxial assist gas to limit oxidation in the process area, enhance oxidation, or promote some other beneficial condition.

Most current lasers have sealed discharge tubes and are either radio frequency (RF) or direct current (DC) excited. Internally, the big difference is that RF is cleaner than DC. The high-voltage DC discharge sputters metal off the electrodes and onto the tube and mirrors. The result is that optics may need to be replaced more often and that metal must be occasionally etched off the tubes. It also breaks down the laser gas, requiring a constant flow of fresh gas to maintain operation. With an RF discharge, the laser cavity can be sealed. The resonator chamber stays clean and the gas can remain in equilibrium.

In terms of performance, one big difference is that the continuous wave (CW) power can be changed with a DC discharge by varying the current. RF lasers are only "on" and "off." This forces the control of power using pulse width modulation. Somewhat mitigating this limitation is the fact that RF can be switched much faster than DC. DC lasers are limited to about 5 kHz, but RF lasers can go over 200 kHz. At 200 kHz, the output is effectively CW.

A lot of the performance differences between DC- and RF-excited CO_2 lasers stem from the differing laser architectures that employ those means of excitation. The original slow-flow DC-excited lasers are limited in average power by heat conduction through the glass tubes. This is why they are so big—a meter of tube is needed for every 50 W of power, but there is a large amount of CO_2 gas in all that tube length, allowing very high peak power to be obtained. Slow-flow DC lasers can produce much higher power pulses than any other kind. Newer DC lasers use pumps to circulate the laser gas so that it can be cooled away from the discharge region. This allows the laser to be more compact, but limits the peak power because there is less CO_2 in the active medium. RF lasers are of the waveguide design or have "tulip" resonators, which have two large plates as electrodes. These minimize the volume of the active medium and thus have relatively low peak powers.

With respect to performance, slow-flow lasers still do the best ceramic scribing because of their high peak power. They can also produce the best surface finish for metal cutting because they have very little noise in their CW output. Fast-axial flow lasers deliver high average power with good beam quality from their stable resonators. RF-excited lasers have the best pulse control and are good when heat input must be limited, such as in the intricate cutting of ceramics.

The CO_2–TEA lasers are used in an imaging configuration because of their high-order multimode beams. This involves imaging a mask on target at some specified demagnification. These lasers are also slow-flow gas lasers, so gas needs to be replenished constantly. Both laser types can, in principle, be used with galvanometers or fixed-beam setups and both can be line narrowed.

2.3 Solid-State Nd^{3+} Lasers

Solid-state lasers are constructed by doping a rare-earth element or metallic element into a variety of host materials. The most common host materials are $Y_3A_{15}O_{12}$ (YAG), $LiYF_4$ (YLF), and YVO_4 (vanadate). The Nd:YAG laser is discussed because it is the most common solid-state laser in industry. These lasers are near-IR lasers, so feature sizes of 20 μm on target are easily achievable using a variety of optical configurations.

2.3.1 Important Characteristics of Nd Lasers

Typical solid-state lasers are pumped optically by arc lamps, flashlamps, or diode lasers (DPSS = diode pumped solid state). Arc lamps are used for CW pumping and flashlamps are used with pulsed lasers. Diode laser pumping is the standard for high repetition rate applications and has opened doors to new industrial applications. Solid-state lasers operate on transitions between different electron energy levels. The atoms of the active medium become excited when an electron jumps to a higher energy level. In Nd lasers, the neodymium ions (3^+) constitute the active medium. Nd lasers are easy to pump. All Nd lasers are four-level systems with numerous energy levels above the upper lasing energy state with quantum designation $^4F_{3/2}$. Atoms in these higher energy states readily decay to the $^4F_{3/2}$ upper lasing state, making it easy to establish the required population inversion.

The emission wavelength of Nd doped lasers varies somewhat with different host materials. Some host materials have a less defined lattice structure than others and as a result the transition wavelengths are slightly different. A very important characteristic is that Nd lasers can be frequency doubled, tripled, or quadrupled through harmonic generation to create shorter wavelength photons—at the cost of some output energy (typically, each harmonic has about half the output energy as the last wavelength). Nd lasers respond well to q-switching, a method described later that emits a pulse train of higher peak power pulses from an otherwise CW output.

2.3.2 Q-Switching

Photons that evolve from spontaneous emission in directions other than along the laser axis are amplified like those along the axis. These photons, however, are not reflected back into the cavity and are lost in the environment. The combined loss of photons traveling off-axis is called amplified spontaneous emission (ASE). A pulse energy enhancement technique called "q-switching" is used to minimize the negative effects of ASE. There are several switching devices in use, both optical and electrical. One such q-switch device is an electronic shutter, sometimes called a Pockels cell, which is triggered open and shut by an electrical signal.

Figure 2.9 shows the q-switch technique, step by step:

1. Laser is pumped with Pockels cell shut; cavity loss is high because the shutter prevents oscillation.
2. Population inversion (gain) grows because pumping continues, but there are few photons to invoke stimulated emission.
3. Pockels cell opens.
4. Cavity loss is greatly reduced now that oscillation is permitted.
5. Optical output is produced, causing population inversion to diminish and gain to reduce.
6. Sequence is repeated for each laser pulse.

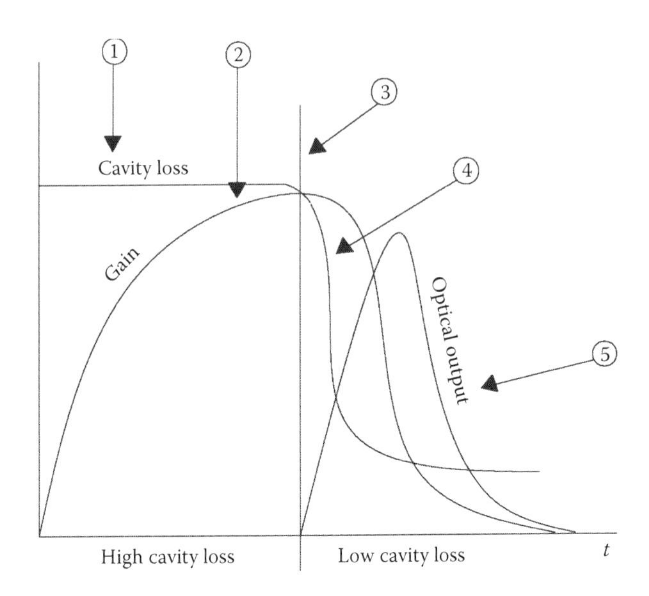

FIGURE 2.9
Q-switching, step by step.

2.3.3 Nd:YLF versus Nd:YAG versus Nd:YVO$_4$

Not all Nd^{3+} lasers have identical characteristics. For instance, the Nd:YAG (YAG = yttrium aluminum garnet) fundamental wavelength is 1.064 μm, and the Nd:YLF (YLF = yttrium lithium fluoride) fundamental wavelength is 1.047 μm. This small difference in wavelength is inconsequential, but there are other differences. Nd:YLF pulse energy is greater than that for a similarly constructed Nd:YAG laser at low pulse rates as more energy can be stored per q-switched pulse because the Nd:YLF upper state energy level lifetime, t_{YLF}, is about three times longer than t_{YAG}. The YAG host material has better thermal conductivity and a more stable refractive index than YLF. As a general statement, because of the long upper state lifetime, the YLF laser has a higher energy per pulse, but limited repetition rate. The vanadate laser (YVO$_4$), on the other hand, can be pulsed at a very high repetition rate, but has less energy available per pulse. The YAG laser is a good compromise between the two. Other solid-state lasers (ruby, glass, and alexandrite) are also available for materials processing.

2.3.4 Harmonic Generation

As a result of a proliferation of laser experimentation in the 1960s, it was discovered that some materials exhibit a nonlinear optical effect when irradiated with high-energy laser emissions. More specifically, the electric dipoles established by the electrons and the nuclei in these materials oscillate in response to incident radiation such that two separate wavelengths of

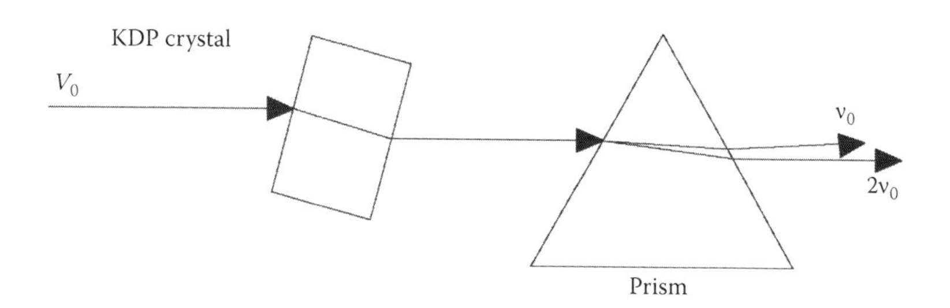

FIGURE 2.10
Harmonic generation of a laser beam through an anisotropic medium.

light exit the material. The output of these emissions includes the original wavelength and a component half the wavelength of the incident beam. This phenomenon (shown in Figure 2.10) is called *harmonic generation*. Other techniques also allow sum and difference frequency light generation.

Harmonic generation is useful in creating different wavelengths; however, total output energy of the shorter wavelength component is typically reduced to one-half the energy of the incident radiation or less. Multiple wavelengths present after the doubling crystal are filtered, to get a single output wavelength, by using prisms or coated optics. Typical doubling crystals include BBO (β-barium borate), KDP (potassium dihydrogen phosphate), KTP (potassium titanyl phosphate), and lithium niobate. These crystals have the necessary properties of birefringence (the decomposition of a ray of light into two rays when it passes through certain materials) and transparency (at both the impinging and exiting wavelengths) and high damage threshold. Organic polymeric materials are being investigated to replace crystals as they are cheaper and can show superior performance.

Second harmonic generation of a Nd:YAG laser results in light output at 532 nm wavelength, in the "green" portion of the visible spectrum. Because many materials are transparent at 532 nm, it is a good wavelength for processing thin films on glass or polymers as there is no damage to the underlying substrate. Green light also couples quite well with some metals, notably copper. The major drawback to using a 532 nm laser is that this wavelength is extremely dangerous to the eye. It transverses through the cornea and is focused by the lens onto the retina, creating a retinal burn, resulting in permanent eye damage; therefore, great care must be taken when using this wavelength.

Third harmonic generation results in light output at 355 nm wavelength—in the UV. This is a good all-around UV wavelength that couples well with many materials, including metals, ceramics, and dielectrics. It is reasonably gentle on the optics and does not require extraordinary safety concerns. Commercial lasers are available from a couple of watts up to a 50 W output.

Fourth harmonic generation results in light output at 266 nm wavelength (also in the UV). This wavelength couples even better than 355 nm, but these lasers are commercially available with only a few watts output and the optics tend to degrade faster, so they require somewhat more care than 355 nm lasers.

2.4 Excimer Lasers

A unique characteristic of the rare or noble gases is that these gas molecules will not normally form compounds with other elements in their ground energy state. The rare gases will combine with certain elements, however, in their excited state. Such a compound is called an "exciplex" molecule and can be used as the active medium in an excimer laser. Table 2.2 shows several different available excimer wavelengths; the examples given discuss KrF (248 nm), but the same discussion applies for all the wavelengths.

2.4.1 Excimer Laser Energy Transitions and Pump Scheme

The electronic pump scheme for the KrF excimer laser is shown in Figure 2.11. The lower Kr + F state is unbound or repulsive; the Kr and F atoms cannot move close to each other because of the lower state energy barrier at the far left. When pumped by the gas discharge, the Kr and F atoms are ionized and

TABLE 2.2

Excimer Gas Mixtures

Mixture	Wavelength (nm)	Gas Life (Pulses)	Average Power (W)	Comments
F_2	157	~10^5	<5	Absorbed by optics and air; requires vacuum beam delivery
ArF	193	~10^6	30	Good for low-power, high-resolution industrial applications
KrCl	222	2×10^6	30	Lower power, short gas life, not very useful
KrF	248	10^7	50–200	Good industrial wavelength and power
XeCl	308	2×10^7	50–200	Good industrial wavelength, particularly on glass products
XeF	351	10^6	<50	Not absorbed by some materials

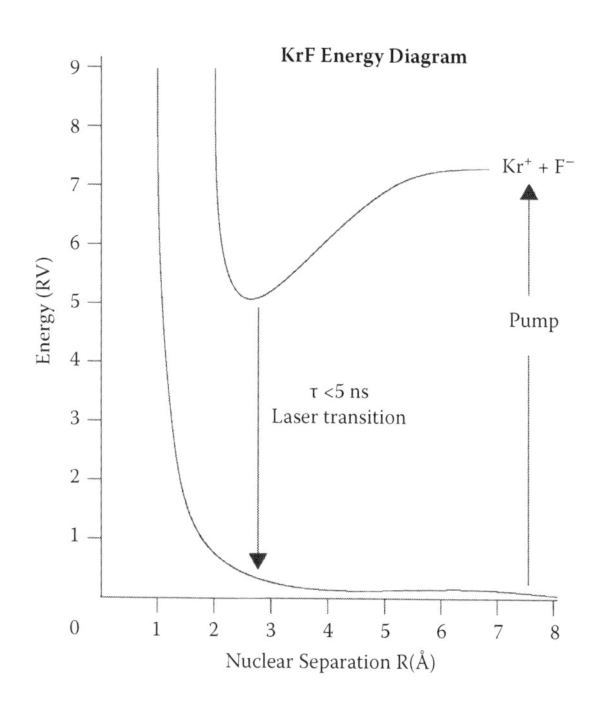

FIGURE 2.11
Energy diagram and pumping scheme for KrF excimer laser.

form the excited dimer molecule at the upper energy state labeled $Kr^+ + F^-$. The atoms can approach closer, now that the previous energy barrier no longer exists. The lifetime for the KrF^* molecule (the * denotes that this molecule only exists in the excited state) in this state is less than 5 ns, during which stimulated emission must occur or the atoms will fall to their ground state spontaneously or through collisions.

Since the lower state does not exist in the bound condition, by simply placing one molecule into the bound upper state, a population inversion is achieved. This condition and the high stimulated emission cross section for KrF^* make a population inversion easy to establish and the laser medium gain very high. The frequency of the excimer UV transmission is sufficiently energetic to break the chemical bonds of many materials and is accomplished through ablation instead of thermal overload. The laser medium in an excimer laser is pumped by a high-speed transverse electrical discharge. DC high voltage is supplied to the pulse-forming network that consists of a thyratron or solid-state switch, magnetic pulse compression circuit, and storage capacitors. When the switch is closed, a high-voltage spike is impressed across the preionization pins and electrodes, ionizing the gas and pumping the excimer atoms to their excited state.

The pumping mechanism for excimer lasers is a gas discharge with up to 45 kV peak excitation voltage. High voltages and currents of this magnitude test the limits of electronic technology. High-voltage power supply failures in excimer lasers are an issue to address in proper laser design. The partial reflectance of the front resonator mirror is only 10% because of the high gain.

A typical excimer laser gas mixture is a $Kr:F_2:Ne$ blend with neon constituting most (98%) of the volume. The neon acts as a third body collision partner in the formation of the excited KrF^* molecule. Excimer lasers are more complex than most other types of UV lasers, require more maintenance, and are more expensive to maintain.

2.4.2 Gas Discharge

The actual gas discharge excitation process takes place in four steps: preionization, kinetic transfer, formation of excited dimers, and laser transition (as illustrated in Figure 2.12).

Typical industrial excimer lasers employ spark preionization to achieve this. The preionization pins are timed to fire just prior to the time when full high voltage reaches the electrodes. When fired, the preionization pins emit a high-intensity electrical pulse that creates a plasma containing a high density of electrons. An initial electron density of 10^7–10^8 electrons/cm^3 is required to produce a sufficient population inversion between the upper and lower energy states:

Step 1. The switch actuates and places up to 45 kV across preionization pins and electrodes, creating a gas plasma.

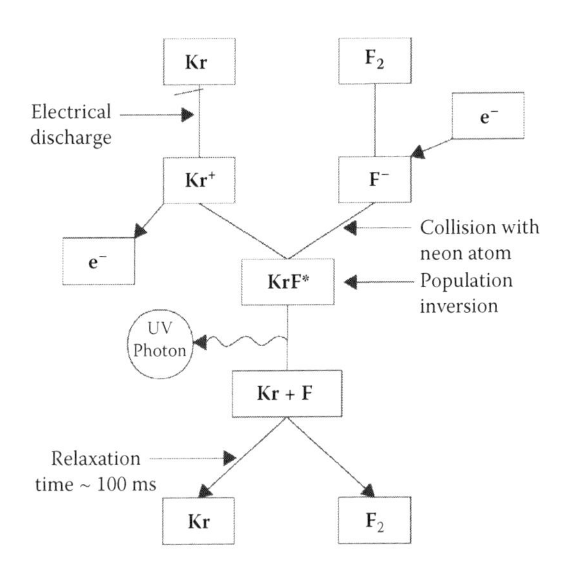

FIGURE 2.12
Simplified diagram of molecular transitions in the KrF excimer laser.

Step 2. The electrons in the gas plasma are accelerated by the electric field between the electrodes as they transfer their kinetic energy to the surrounding atoms.

Step 3. Excited KrF^* molecules are created by inelastic collision with the electrons. These molecules have an approximate lifetime of >5 ns in their excited state and will decay spontaneously if not stimulated by an additional photon.

Step 4. The laser transition step is initiated by those photons produced by spontaneous emission along the laser axis. These photons are reflected back along the axis by the resonator optics at each end of the laser so that they may subsequently produce stimulated emission with other excited molecules. The laser emission occurs in about a 20 ns pulse because the electronic circuitry cannot sustain a constant high voltage and the gas discharge is short-lived.

After the pulse is completed, the gas constituents require a 100 ms relaxation period before they can participate in the next discharge cycle.

The relaxation time requirement of the excimer gas mixture places a significant constraint on the pulse rate capability of the laser. The laser is limited to a pulse rate compatible with the 100 ms relaxation time unless the gas between the electrodes is replenished. A typical excimer laser overcomes this constraint by recirculating the laser gas so that the volume of the gas is completely refreshed and exchanged several times between laser pulses. At the same time, the gas is cooled and filtered during the circulation process such that repetition rates of 400 pulses per second or higher are achievable. However, in no case can the repetition rate be higher than a couple of kilohertz, and these high repetition rate waveguide lasers have lower per-pulse energy.

2.4.3 Excimer Laser Energy Monitoring

During normal operation, laser output energy depends on the high-voltage set point. Raising this set point increases the energy of the laser beam. As the gas ages, the output energy for a particular high-voltage setting will decrease. Therefore, the output energy of the excimer laser must be monitored to ensure uniform machining. An energy monitor mounted at the front aperture provides pulse energy information to the laser control computer. As the energy decreases, the computer increases the high-voltage set point to compensate. The voltage can also be adjusted manually. Eventually, the high-voltage power supply limit will be reached and separate measures must be undertaken. Three adjustments to the gas mixture are possible:

- *Halogen injection.* A spurt of fluorine gas injected into the laser reservoir can extend gas lifetime significantly. This method can be repeated several times until further injections are ineffective.
- *Partial gas replenishment.* A portion of the laser gas is removed and replaced with fresh volume in the correct component gas ratios. However, at some point, even this method becomes ineffective.
- *New fill.* The entire reservoir is evacuated and replaced with fresh gas. A new fill is necessary when other methods fail. Note that excimer lasers have both a static and a dynamic gas lifetime. The gas lifetime is usually stated in "number of shots to 50% power." However, this assumes a fresh fill with new optics, and it should be further noted that if a fresh fill is placed in the laser but is not used for days or weeks, the gas is probably no longer useful. The gas lifetime of a 193 nm laser is less than 248 nm < 308 nm.

After a new gas fill, the excimer laser beam profile is Gaussian on the short axis and flat topped along the long axis, as in Figure 2.13(a).

As the gas ages, changes in chemistry alter its electrical properties. These changes result in beam growth along the short axis. The beam changes to a flat-topped profile on both axes, beam divergence increases, and peak pulse energy is reduced in the far field (Figure 2.13b). A halogen injection or partial gas replenishment will counteract these changes temporarily, but full beam profile will not be restored completely without a new gas fill.

Sputtered electrode and preionization pin material over the life of the laser generate minute particles of dust within the resonator cavity. The dust precipitates out on the resonator windows as well as on other internal components. Dust on the windows absorbs UV light, which reduces the energy output of the laser and, over time, can damage the cavity optics, particularly in the central portion (Figure 2.13c). Therefore, regular window cleaning is critical to proper laser operation.

Resonator optics must be aligned perpendicularly to the beam axis for efficient laser oscillation. Misalignment of the windows causes the optical feedback to be skewed with respect to the gain medium, resulting in lower overall gain. Misalignment in the long axis results in hotspots in the beam profile; misalignment along the short axis causes a dramatic drop in total power, often without discernible hot spots (Figure 2.13d).

As the preionization pins wear, the spark gap increases, adversely affecting gas preionization. As the cathode wears and becomes flatter, the discharge becomes nonuniform. Worn electrodes or preionization pins can cause a trapezoidal or split beam, as shown in Figure 2.13(e). This effect cannot be corrected by resonator alignment and becomes more pronounced as the gas fill ages. The resulting nonuniform power density of the beam is unacceptable for critical applications. Laser refurbishment is the only remedy.

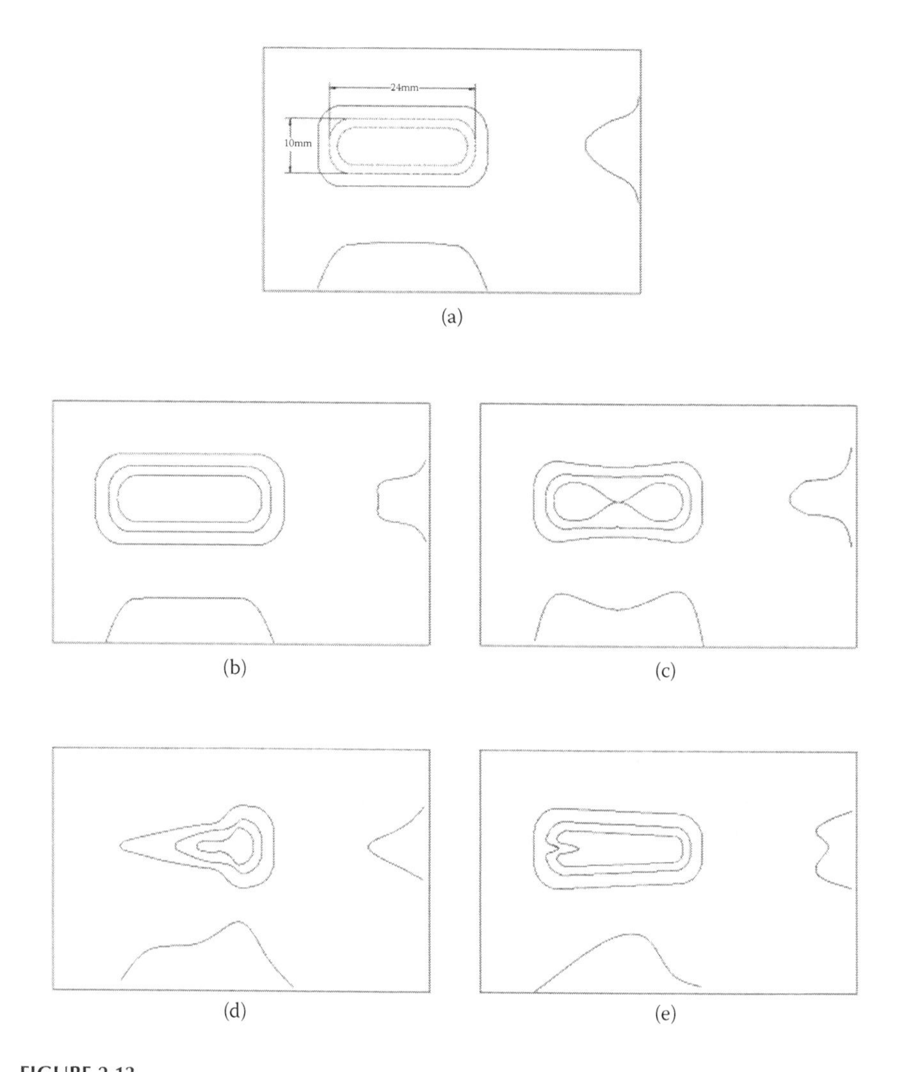

FIGURE 2.13
(a) Normal beam profile with new gas fill (top); (b) beam profile of an old gas fill (middle left); (c) beam profile of a laser with dirty optics (middle right); (d) beam profile of a laser with misaligned resonator optics (bottom left); (e) beam profile of a laser with worn electrodes or preionization pins (bottom right).

Excimer lasers were the first method developed for generating UV photons and, prior to about 1995, were the only industrial UV laser. UV photons can now be obtained from several other different laser sources, but excimer lasers, despite their drawbacks, still can do things no other lasers can do and, in some cases (a lot of small holes on close centers), they can be economic since they use mask imaging instead of single point drilling. The resonator cavity configuration produces a beam ideal for near-field imaging and the high peak power of the laser beam permits ablation of the target material with little or no heat affected zone. The 193–351 nm optical wavelength range permits generation of high-resolution (~1 μm) features on the target surface, and the shallow absorption depth permits tight control of feature depth by controlling the number of pulses applied. The large beam cross section accommodates a large imaging mask for near-field imaging.

High-performance electronic components in the laser require frequent and costly maintenance. Laser gas is toxic, corrosive, and expensive and laser gas consumption is high. Changes in gas chemistry affect beam shape and quality, and components inside the laser require routine replacement and cleaning due to corrosiveness of the laser gas. Resonator optics and beam delivery optics degrade with the exposure to UV light and require replacement, and optics need routine cleaning.

2.4.4 Operation and Maintenance Costs

Operation and maintenance of an excimer laser materials processing system typically requires the attention of one full-time operator and a part-time maintenance technician (3–5 hours/week). Excimer lasers are designed to operate over a range of different pulse rates. Lasers at high pulse rates typically have lower output power. Therefore, a laser running at 200 pulses per second continuously would still require refurbishment after two billion pulses, which would occur after 6 months of continuous use. A refurbishment normally includes cleaning and acid etching all internal surfaces, electrode replacement, complete preionization pin replacement and fan assembly, and precipitator replacement. The refurbishment may also include the replacement cost of the high-voltage switch, heat exchangers, capacitors, halogen filter, or resonator optics.

2.5 Fiber Lasers

Fiber lasers were first conceived in the early 1960s, but their utility was not taken advantage of until the late 1990s when a vast amount of money was put into development for the communications industry. Although fiber lasers were never really adopted wholesale in telecom systems, the real benefit came from development, cost reduction, and robustness of the contributing components; eventually people realized that these lasers could be used for material processing and not just for data transportation.

Fiber as a gain medium matured during the telecom decade in the form of amplifiers critical to boosting data signals for transmission over hundreds of kilometers of fiber. Investment in fiber amplifiers created robust and inexpensive component technology that enabled them to mature as industrial laser sources over the past decade. Key component technologies

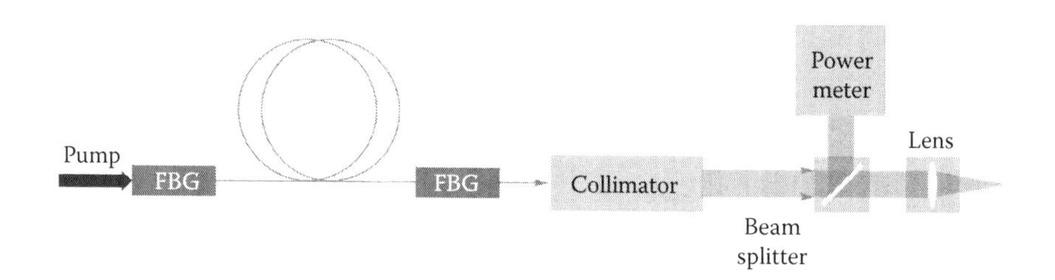

FIGURE 2.14
Simple fiber laser diagram.

from the telecom industry include high-power, reliable pump diodes; environmentally robust single-mode fiber packaging; mature production of doped glass fiber to high tolerances; and passive fiber optic components to monitor and control optical performance. Figure 2.14 shows a simple fiber laser pumping diagram.

In a fiber laser, the active gain medium is an optical fiber doped with rare-earth elements such as erbium, ytterbium, or neodymium. The advantages of fiber lasers over other types include the following:

- *Light is already coupled into a flexible fiber.* This allows it to be delivered easily to a movable focusing element. This has important benefits for laser cutting, welding, and drilling of metals and polymers.
- *High output power.* Fiber lasers can have active regions several kilometers long, so they can provide very high optical gain. They can support kilowatt levels of continuous output power because of the fiber's high surface area to volume ratio, which allows efficient cooling.
- *High optical quality.* The fiber's wave-guiding properties reduce or eliminate thermal distortion of the optical path, typically producing a diffraction-limited, high-quality optical beam. Since the optical mode is defined by a fixed solid waveguide, the beam quality is maintained without need for adjustment or tuning over the life of the laser, and it is invariant over a wide range of operating conditions. This eliminates the need and cost for beam characterization after manufacture.
- *Compact size.* Fiber lasers are compact compared to rod or gas lasers of comparable power because the fiber can be bent and coiled to save space.
- *Reliability.* Fiber lasers exhibit high vibrational stability, extended lifetime, and maintenance-free turnkey operation.
- *Cost.* These lasers are almost cost competitive with CO_2 lasers to buy (\$/W) and also to operate at lower power levels (<200 W); at higher power levels, the cost curves cross.

Fiber lasers employ a special design concept known as "double cladding," which can be thought of as a "waveguide within a waveguide." A tight waveguide in the center ("core") of the fiber contains a rare-earth dopant, while the glass cladding for that core forms a highly tolerant waveguide for low brightness pump light. The cladding glass is covered by a flexible acrylate or polyimide coating that serves as the lower index "pump cladding" and offers physical protection to the glass fiber from moisture and mechanical damage.

The fundamental wavelength of fiber lasers is between 1 and 2 μm depending on the dopant and host material. Fiber lasers are available from a few watts to multiple kilowatt output, but micromachining applications will typically use lasers with output of less than 200 W. These lasers can also be q-switched, so short, high-peak power pulses can be obtained at pulse repetition rates of tens or hundreds of kilohertz. The fundamental wavelength couples extremely well with metals, and fiber lasers have replaced other solid-state and CO_2 lasers in many applications. The significant advantages discussed before make fiber lasers the fastest growing industrial laser over the past decade, now comprising over 25% of the global industrial laser market.

Fiber optic components and double-clad pumping allow for design innovations that directly enhance the material processing capabilities of fiber lasers. Some examples include the following:

- *Control and stability of output power.* By tapping minute but precise samples of the power in the fiber and controlling current to the pump diodes, fiber lasers can operate in "closed loop" mode with power stability better than ±0.5%. The same tap monitoring can be used to maintain absolute power calibration over life. This gives material processing applications a degree of certainty and repeatability that is only possible with time-consuming and expensive procedures when other laser sources are used. The stability of the laser is enabling for highly sensitive processes (such as cutting and welding medical devices or in smart phone production) and provides high yield and throughput for low-cost, high-volume, high-precision processes.
- *Alternatives to q-switching, such as MOPA (master oscillator, power amplifier) design.* These allow for nearly independent control of pulse width, pulse repetition frequency, and average power. A common MOPA design uses a tiny fiber-coupled semiconductor diode laser as a seed oscillator. Diode laser sources capable of short-pulse (down to 1 ns) and high-repetition frequency (over 1 MHz) operation were developed for telecom. Adapted to match the gain of optical fibers, output from these seed sources can be changed or tuned almost instantaneously and the output amplified to industrial power levels by subsequent single-pass fiber amplifiers. This operational flexibility allows a single pulsed fiber laser to perform a multitude of micromachining applications. Historically, laser process development has

involved designing a process around the limitations of the laser; with MOPA technology, simple software commands make it possible to "tune" the laser operation to optimize the desired process effect. One example is a deep engrave and polish of metal, whereby a low-repetition-rate, high-peak-power setting delivers high-pulse energy for fast material removal; then, a quick change in the program switches the laser into a high-repetition rate, low-pulse-energy state to smooth the etched surface rapidly.

- *Precise and stable low-mode beam design.* Many laser designs are most stable as either single-mode (Gaussian) or highly multimode resonators, providing the optics designer with either a sharp Gaussian beam or a broad, "flat-top" profile. Low-order Gaussian mode beams are normally used and can be focused to the smallest spot size with a large depth of field; however, the central hot spot can be problematic. Flat-top beams lack the peak intensity, depth of field, and small spot of Gaussian beams, but provide a uniform intensity—and therefore processing effect—on target. When preferred, fiber lasers can be engineered so that the M^2 value is between 1.5 and 5, thereby producing larger, softer beam profiles (with increasing M^2) while still maintaining high intensity and depth of field. Applications that take advantage of this characteristic are single line fonts readable by the naked eye and microwelding of thin metals.

2.6 Disk Lasers

A disk or active mirror laser is another type of doped solid-state laser. Disk lasers have a thin layer of substrate with an active gain medium. There is a heat sink on the opposite side of the laser output (Figure 2.15). Disk lasers do not have to be circular; other shapes have also been tried. Disk lasers can be thought of as a special variant of slab lasers, with the benefit of more efficient heat removal across the thin direction and the ability to scale power by increasing area while maintaining uniform material properties and even optical excitation. They can be scaled to very high powers, but low-power lasers do offer some benefits in micromachining, such as excellent beam stability, low divergence, and high beam quality. Pulse lengths in the range of a couple of nanoseconds can be obtained.

The most common active medium is Yb:YAG, but many others, including Nd:YAG, are also available. There is currently a very big competition between disk and fiber lasers (mostly in metal-processing applications) and these two sources dominate other lasers sources in the 2–20 kW continuous wave power range. Disk lasers are used in the medical industry to process

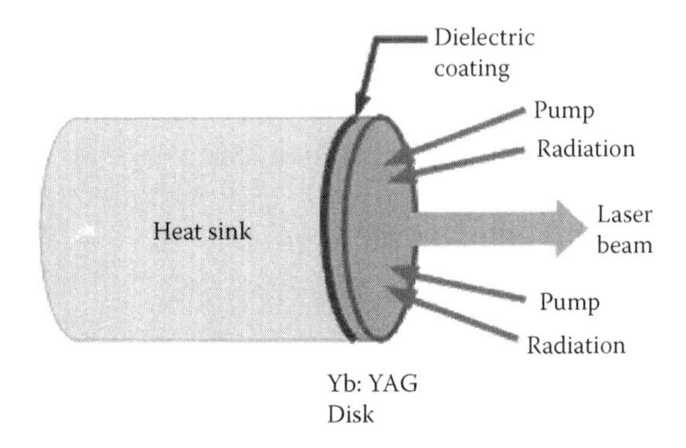

FIGURE 2.15
Simple disk laser diagram.

stainless steel, nitinol, tungsten, platinum and their alloys as they tend to leave less slag than other IR lasers. Frequency-doubled 515 nm lasers are also available and the third harmonic is being developed.

2.7 Ultrashort Pulse (USP) Lasers

In this text, USP lasers are defined as having a pulse length less than 1 ns, including picosecond (10^{-12} s) and femtosecond (10^{-15} s) lasers. Because of the very short pulse length, extremely high peak powers can be reached, resulting in excellent edge quality. These lasers typically run up to 100 kHz repetition rate or more. To put things into perspective, consider a laser with a 1 ps output operating at 100 kHz repetition rate. This would be like closing your eyes, blinking for 1 second, and then keeping your eyes closed for *4 months.* So, the on-time or duty cycle of these lasers is extremely small.

Table 2.3 compares the pulse lengths of commercially available lasers. An interesting phenomenon of USP lasers is that at *very* short pulse lengths,

TABLE 2.3
Pulse Lengths of Commercially Available Lasers

Laser	Pulse Length
CO_2	50–150 μsec
CO_2–TEA	200 ns
Pulsed Nd:YAG	1–1000 μs
DPSS laser	20–100 ns
Q-switch fiber laser	100–500 ns
Short pulse laser	100 fs–1 ns

wavelength-dependent absorption appears to be eliminated due to multi-photon effects and very high absorptions. Clean material processing can be achieved using the fundamental IR wavelength. USP lasers, when used properly, can give virtually no HAZ (heat-affected zone) or microcracking.

The exact mechanisms of material interaction are complex and the details are still debated; in summary, short pulse lasers have pulse lengths shorter than the reaction time of the material lattice, and therefore the photons are absorbed by free valence electrons resulting in bond breaking and material ejection before heat can be transferred into the material. Longer pulse lasers tend to melt a pool of material at the interaction zone, which then reacts with photons at the tail of the pulse to eject and evaporate molten material. This is accompanied by splashing and melting and, because of the longer pulse length, heat transfer into the material.

Picosecond lasers are available in multiple wavelengths because they have higher energy per pulse than femtosecond lasers. In some materials, there is no identifiable difference between the fundamental, doubled, or tripled wavelengths, so it is better to use the higher energy available in the fundamental. However, many materials do show wavelength dependence in the picosecond regime, so there may be reasons to choose one particular wavelength over another. Additionally, basic physics still applies, and smaller spot sizes are achievable with shorter wavelength light. Where does true wavelength independence occur? Probably one needs to be below 100 fs pulse length.

In general, one would want the shortest pulse length possible for the best micromachining results, but there are practical considerations. As the pulse length of a laser gets shorter, the available pulse energy is usually smaller. More pulse energy can therefore be obtained from a picosecond laser than from a femtosecond laser. Also, in general, the lasers become more expensive (especially on a dollar-per-watt basis) and more difficult to use as the pulse length gets shorter. When deciding on the proper laser to use, one must take into account all considerations, including cost and speed. In general, for *most* current applications, longer pulse lasers are "good enough." Short pulse lasers are therefore used when quality concerns or other considerations outweigh cost concerns.

2.8 Comparisons of Laser Sources

Throughout this text a shape similar to those presented in Figure 2.16(a–i) will be used to illustrate the comparisons between different laser types on different materials. These photos show a section of a part about the size of a quarter and containing this noticeable feature with two distinct lines meeting. The kerf width in all cases is about 75 μm and, unless otherwise noted, all photos are taken directly from the laser with no cleaning or postlaser processing. Two of the most common polymers are Kapton™ (polyimide), shown in Figure 2.16, and Mylar™ (polyethylene terephthalate or PET), shown in Figure 2.17; both are DuPont products.

The broadband CO_2 laser in Figures 2.16(a) and 2.17(a) shows signs of burning and melting around the edges in both materials, but especially in PET. The shorter pulse length and higher pulse energy of the TEA laser (Figures 2.16b and 2.17b) give improvement in the edge quality. The fundamental YAG laser

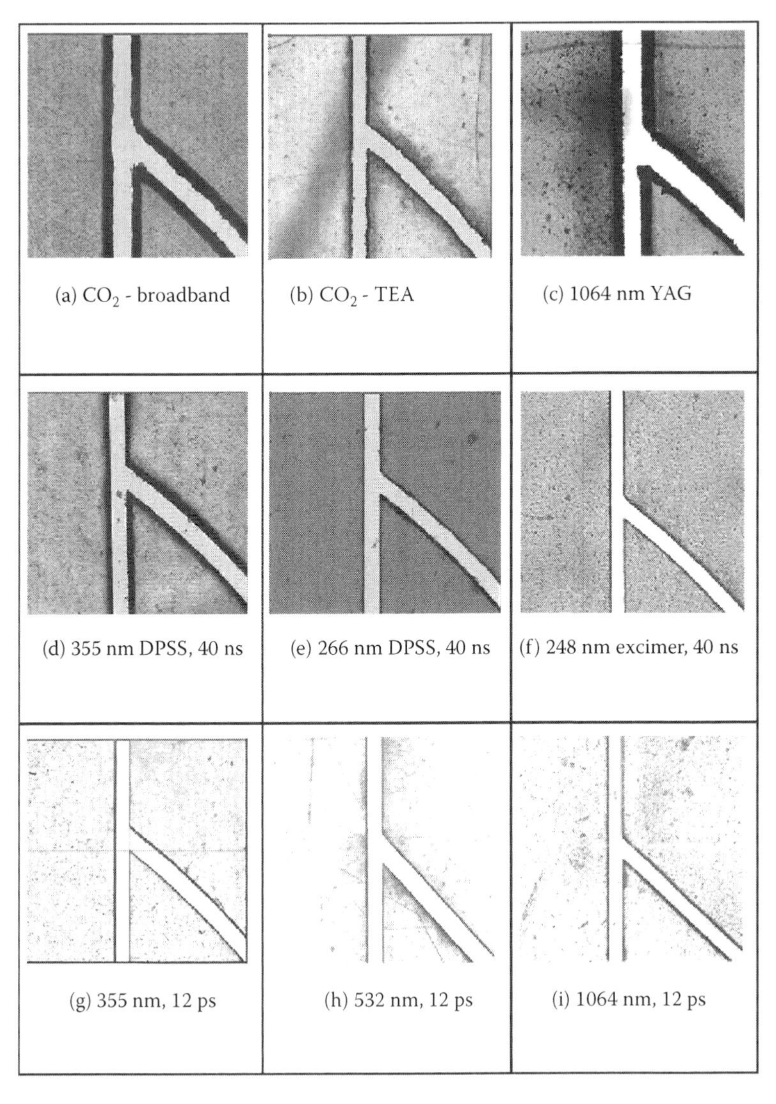

FIGURE 2.16
(a–i) Polyimide processed with different lasers.

(Figures 2.16c and 2.17c) also shows signs of charring and scattered debris. These long-pulse IR lasers all show signs of heat effects. The UV lasers shown in (Figures 2.16d–f and 2.17d–f) show a definite improvement in the edge quality, with the 248 nm excimer laser looking almost pristine, especially on the polyimide. This row of pictures shows the edge quality improvement going to shorter wavelengths. In the final row (Figures 2.16g–i and 2.17g–i), all of the work was done with a 12 ps pulse length laser. Although the fundamental wavelength work is not as good as that which was doubled or tripled, all of these photos show marked improvement over the other photos done with longer pulse length lasers.

Figure 2.18 shows the difference between 355 and 266 nm processing in an otherwise similar laser setup with the same pulse length. Two materials are shown: stainless steel and alumina. It is clear that the shorter wavelength

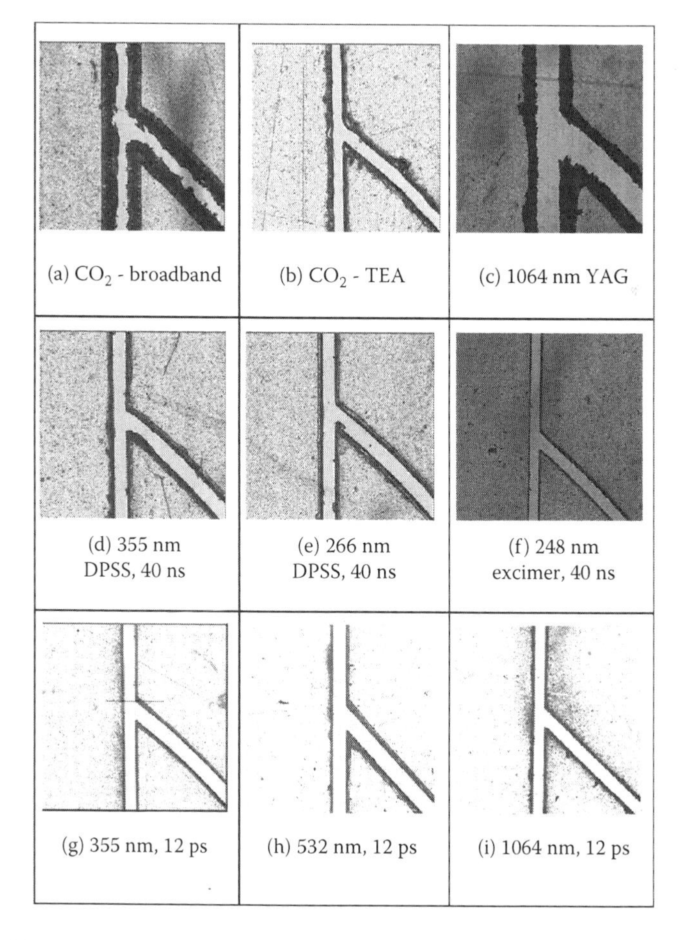

FIGURE 2.17
(a–i) PET processed with different lasers.

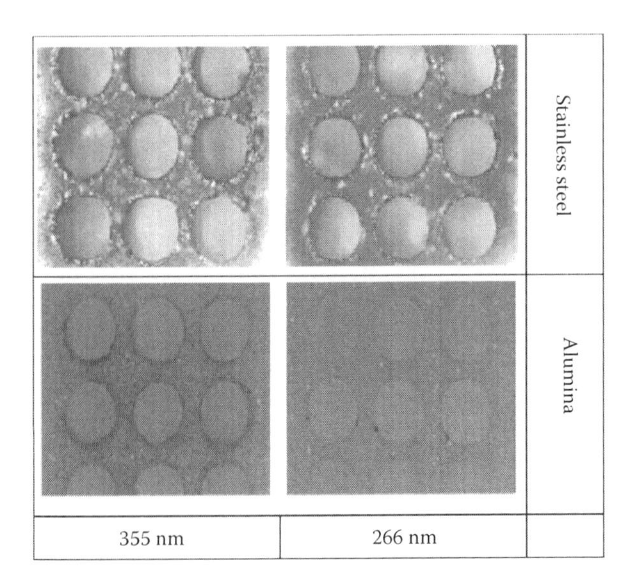

FIGURE 2.18
Stainless steel and alumina processed with 355 and 266 nm lasers.

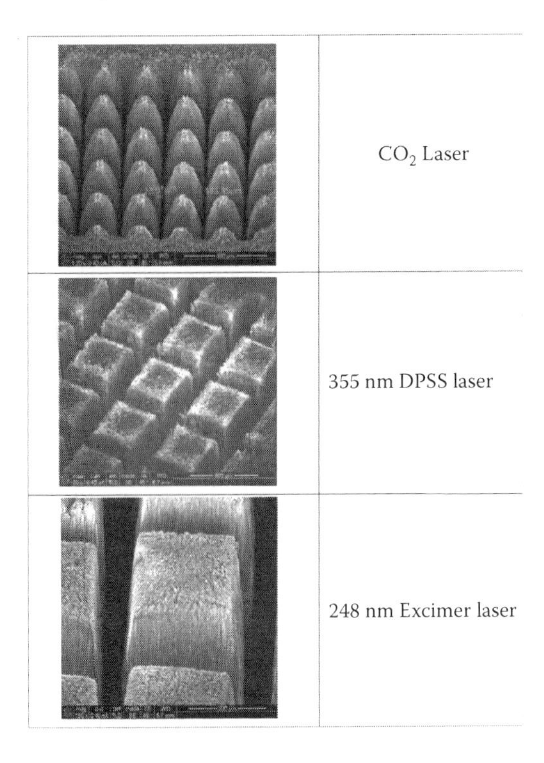

FIGURE 2.19
Laser-cutting comparisons in carbon-based substrate.

TABLE 2.4

Laser Processing Speed and Cost

Laser	Wavelength	Processing Speed	Operating Cost ($/hour)	Capital Cost ($/W)
CO_2	10 μm	<1 second	10	250
Excimer	248 nm	>1 hour	50	4,000
DPSS	1064 nm	<1 second	15	1,000
	355 nm	18 seconds	15	10,000
	266 nm	15 seconds	15	75,000
Fiber	1064 nm	<1 second	10	600

gives better edge quality in both cases. Figure 2.19 shows processing results on a carbon-based substrate using three different lasers. There is clear melt and rounding of the edges when using the CO_2 laser. The 355 nm laser produces reasonably nice edges but nowhere near the quality seen with the 248 nm excimer laser.

So, if the edge quality is best on short pulse and/or short wavelength lasers, why not use them everywhere? There are several reasons. First, the capital cost to *buy* the lasers is higher on a dollar-per watt-basis. Second, the operating costs are also much higher—especially with the excimer laser (Table 2.4). Finally, at least for the excimer laser, the time it takes to make a complete part is much longer than for any of the other lasers because the excimer is not best utilized in "cutting" applications due to its low pulse repetition rate.

However, even using the shortest pulse laser and the shortest wavelength laser available does not guarantee getting the best edge quality on target. If the lasers are used incorrectly or the optics set up incorrectly, poor quality will result.

Now that laser sources have been reviewed, it is appropriate to discuss optics in the next chapter.

3

Optics

3.1 Optics

The part of a laser processing system that directs the beam to the target after it leaves the laser head is the beam delivery system. Its components include masks, turning mirrors, attenuators, field lenses, focusing lenses, scanning lenses, scanning mirrors, beam splitters, and imaging lenses that manipulate and shape the beam, and the optical chamber that holds these devices. Other devices are used to measure a beam's output power and beam profile.

3.1.1 The Law of Refraction (Snell's Law)

In optics, Snell's law is a formula used to describe the relationship between the angles of incidence and refraction when referring to light passing through a boundary between two different isotropic media, such as air and glass. Figure 3.1(a) shows light reflected from a surface in which the incident light, a line normal to the surface, and the reflected light lie in the same plane. Light is reflected at an angle equal to the incident angle. For a light ray traveling through two media, 1 and 2, as shown in Figure 3.1(b), the following relationship is always constant:

$$\frac{\sin\theta_1}{\sin\theta_2} = \text{constant} \tag{3.1}$$

If material 1 is a vacuum, then the constant is specific to material 2 and is called the *index of refraction*. The index of refraction of a material is an indicator of the speed of light traveling in the medium. A more familiar way of writing Snell's law for two materials is

$$n_1 \sin\theta_1 = n_2 \sin\theta_2 \tag{3.2}$$

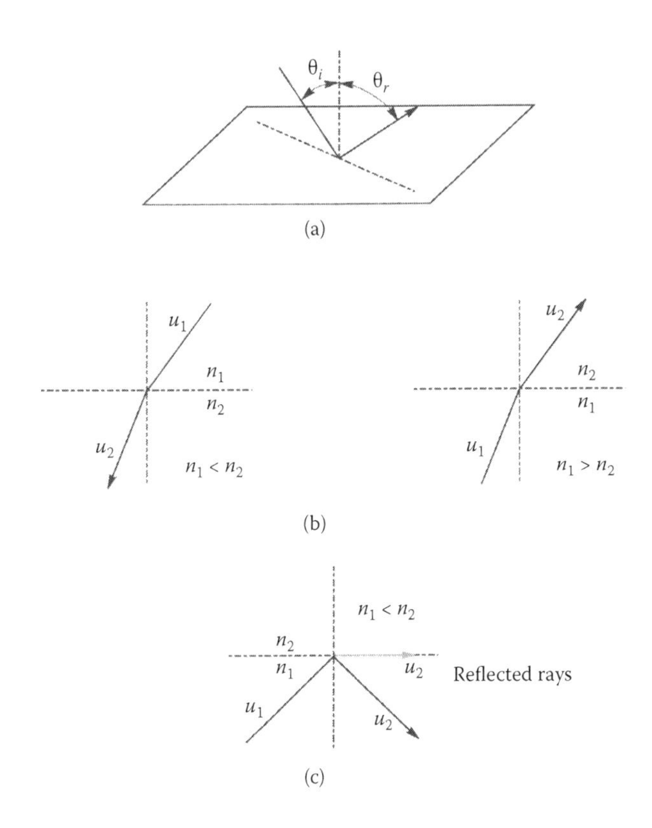

FIGURE 3.1
(a–c) Snell's law.

When light travels from a medium with a higher refractive index to one with a lower refractive index, Snell's law seems to require, in some cases (whenever the angle of incidence is large enough), that the sine of the angle of refraction be greater than one. This, of course, is impossible, and the light in such cases is completely reflected by the boundary, a phenomenon known as *total internal reflection*. The largest possible angle of incidence that still results in a refracted ray is called the *critical angle*; in this case, the refracted ray travels along the boundary between the two media (Figure 3.1c).

3.1.2 Simple Optics—Materials, Substrates, Coatings, Lenses, and Prisms

Materials must have close to 100% transmission at a particular wavelength for use as refractive optics; otherwise, significant energy is lost. Likewise, reflective optics must be close to 100% reflective (either the substrate itself or an applied coating). Sometimes coatings can enhance the utility of simple optics. The price of optical components depends on the substrate material, coating, diameter, focal length, and optical quality. Some considerations in choosing optics include identifying the proper wavelength, compatibility with the surrounding environment (i.e., moisture, temperature), resistance to irradiation, beam size, demagnification and fluence requirements, and cost.

Some optics (mostly those with a high index of refraction) come with an antireflection (AR) or interference filter coating. This helps increase transmission in refractive optics. High-reflection (HR) coatings that decrease losses in reflective optics are also available. Coated optics cost more but may improve machining quality. Consistency of the coatings is critical and is highly dependent on the manufacturer. Flatness is also critical, especially for thin optics. Some ultraviolet optical materials include the following:

magnesium fluoride (MgF_2)—best resistance to radiation damage but most expensive

fused silica (SiO_2)—medium cost, low index of refraction (n = 1.46), reduced transmission at shorter wavelengths, and somewhat difficult to machine

calcium fluoride (CaF_2)—degrades slowly with exposure to electromagnetic radiation, least expensive

Infrared optical materials include the following:

germanium (Ge)—high index of refraction (n = 4.0), exhibits poor transmission qualities at high temperatures (>200°C), and allows no pointer laser transmission

zinc selenide (ZnSe)—a refractive index of n = 2.4, scratches easily, and requires AR coating

sodium chloride (NaCl)—low refractive index, water soluble, and good transmission

High-power fixed-beam Nd:YAG optical components are usually made from fused silica. Scan lenses are typically not made of fused silica since the small index of refraction leads to more expensive designs or worse imaging quality than when other materials are used. Therefore, fused silica is only used in applications requiring high/critical peak powers/energies (to avoid destruction) and high average powers (to avoid thermal lensing, which is an effect seen mostly in high-power lasers where heating of the crystal affects the refractive index and therefore changes the beam quality).

The simplest optic is the positive (converging) *spherical* lens shown in Figure 3.2. This type of lens takes a collimated beam (seen coming from the left) and focuses that beam in all directions to a focal point at a distance, f, from the lens that is called the *focal distance* of the lens. Gaussian beam lasers are normally used in a focal point configuration. This same lens can be used for imaging, where an image is projected in the image plane (beyond the focal point). Note that the original is inverted in the image plane.

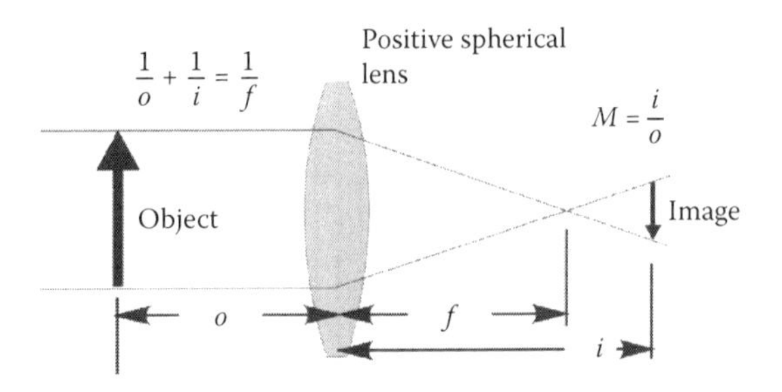

FIGURE 3.2
Positive spherical lens.

Positive lenses focus collimated, incoming light, and negative lenses disperse it such that a "virtual" focus is formed on the input side of the lens (Figure 3.3). Negative lenses are therefore used to disperse collimated light. Because no lens is perfect, there are distortions from the ideal models presented in these diagrams; for instance, the focal point is in theory really a point in space, but in reality there is some depth of field and width associated with the focus. Therefore, multielement lenses are designed so that aberrations can be minimized. They are more expensive, but they allow smaller spots and larger fields on target. An alternative is an *aspheric* lens, which has a more complex surface profile than the simple spherical lens. This profile can reduce or eliminate spherical aberrations and can replace multielement lenses, resulting in a smaller, lighter, and possibly cheaper solution.

A couple other terms are frequently used when describing optics and optical systems. The first is *f-number* or $f^{\#}$. The f-number is defined as the ratio of the focal length to the diameter of the entrance pupil ($f^{\#} = f/D$). The second is *numerical aperture* or NA. Numerical aperture is defined as the index of

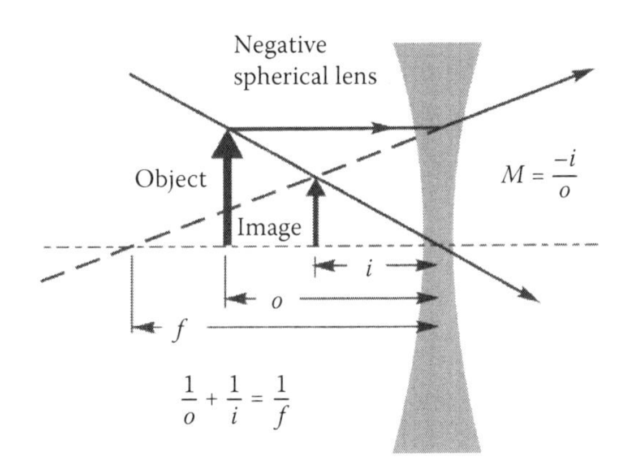

FIGURE 3.3
Negative spherical lens.

refraction of the material times the half-angle of the maximum cone that can be made by light entering or exiting the system (draw a line defining the focal point and take the half-angle); mathematically NA = n sinθ. Assuming that the optic is used in air ($n = 1$) and the numerical aperture is small, NA is approximately related to $f^{\#}$ by the equation NA ~ $1/2f^{\#}$.

A *cylinder* lens focuses light in one dimension only. These lenses can transform a point of light into a line image and they also come in both positive and negative varieties depending on whether the beam is to be focused or dispersed. Figure 3.4(a) shows the optical parameters of a positive cylinder lens and Figure 3.4(b) shows the parameters of a negative cylinder lens.

A *prism* is a transparent optical element with flat, polished surfaces that refracts light as it passes through. Each wavelength is refracted at a different angle (Figure 3.5a), so prisms can be used to separate different incoming wavelengths of light as well as, by using reflections off the surfaces, to split beams (Figure 3.5b). Rotating pairs of prisms called *Risley prisms* (Figure 3.5c) can also be used as automated beam delivery for drilling circular holes either by trepanning or by percussion drilling. Some of the benefits include compact size, insensitivity to vibrations, low moment of inertia, robustness, and speed.

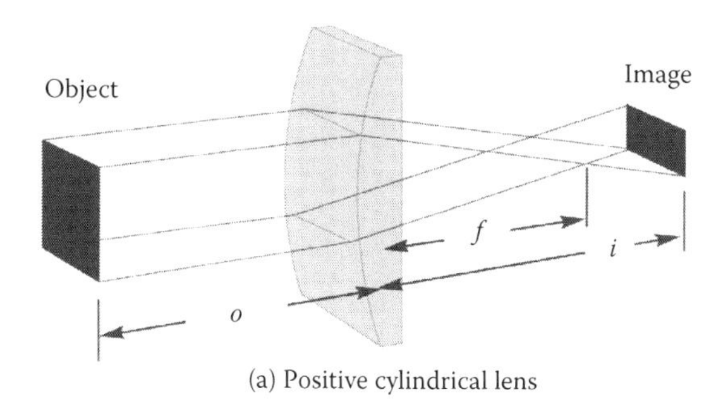

(a) Positive cylindrical lens

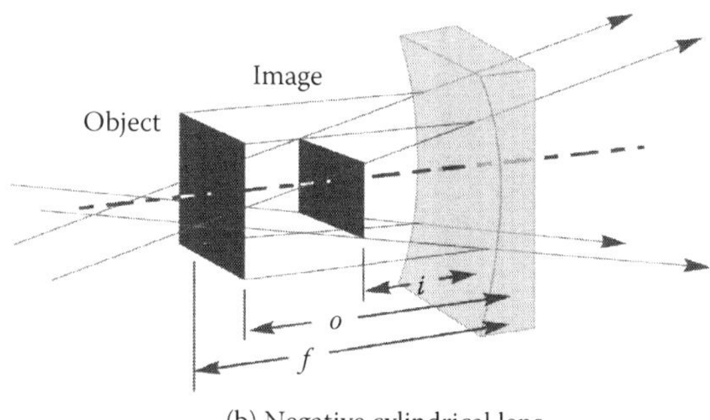

(b) Negative cylindrical lens

FIGURE 3.4
(a) Positive and (b) negative cylinder lenses.

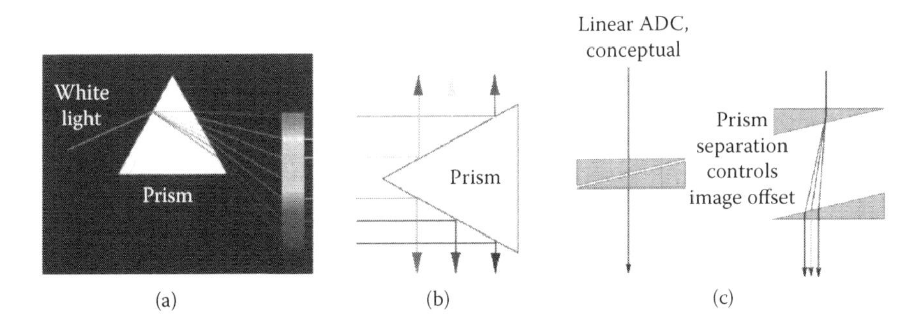

FIGURE 3.5
(a–c) Prisms.

3.1.3 Beam Splitters

Two main methods are used to split an incident beam into two or more components. The first is by using a dielectric coated optic. Depending on the coating, a portion of the beam will be transmitted and a portion will be reflected. For instance, it is very hard to achieve a split in equal halves using a 50%, 45° to incident beam splitter, so in order to keep each beamlet at the same energy, attenuators in each beam line must frequently be used. Dielectric beam splitters transfer the original beam size to both resulting beamlets (Figure 3.6), but they reduce energy density and peak power by half (in a 50/50 split). Thus, the technique has some drawbacks—especially when the beam is to be split into many components.

For a large, multimode beam like an excimer laser, a better way to split the beam is by using a physical divider and "scraping" off portions of the beam (Figure 3.7). By using different portions of the beam, the initial energy density and peak power are maintained in the resulting beamlets. This technique has been used to split an incident excimer laser beam into 12 equal components for parallel processing of 12 parts simultaneously.

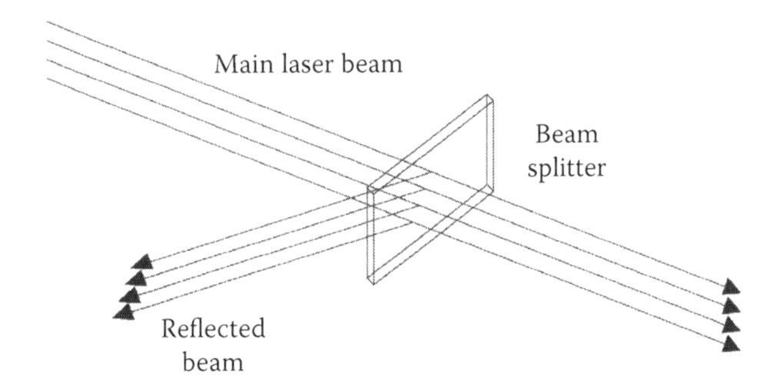

FIGURE 3.6
Dielectric beam splitter.

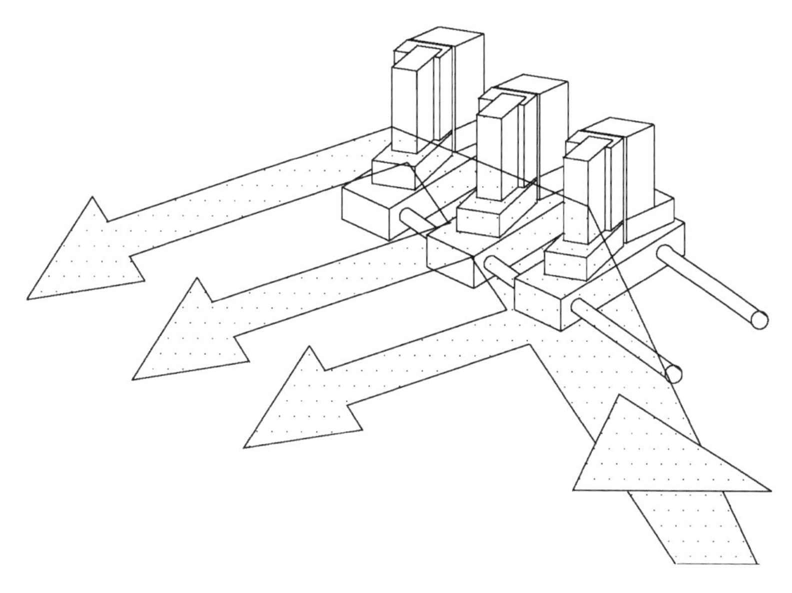

FIGURE 3.7
Physical beam splitter.

3.1.4 Telescopes

Normally, Gaussian laser beams have small diameters. If larger objects are to be illuminated or if one wishes to fill an imaging lens to get the smallest spot size, the beam must be expanded. Expansion can be achieved in two ways (Figure 3.8).

The first method uses two positive imaging lenses of the Kepler telescope type (Figure 3.9). It offers the advantages of having a real focus at which a spatial filter can be easily inserted and it uses inexpensive optics. The disadvantages, however, are that, for powerful lasers, the extremely high beam intensity at the focus can cause breakdowns in air. In addition, the path length needed is much longer since the distance between lenses is the sum of the foci.

The second method uses a negative entrance lens and a positive collimator lens of the reversed Galilean type (Figure 3.10). It offers the advantage of being shorter than the Kepler type and has a virtual focus, which prevents air breakdown. On the other hand, a spatial filter cannot be inserted and this type of telescope is a bit more expensive.

The expansion ratio in each case is characterized by the focal length ratio and the divergence of the laser beam is reduced by the absolute value of the ratio of the two focal lengths. It should be observed that telescopes work in *both* directions: One can expand or reduce the beam size by orienting the telescope in one direction or another. If we wish to reduce the beam diameter by reversing the lens arrangement, the divergence will increase by the absolute value of the focal length ratio (i.e., we can no longer expect "parallel" laser

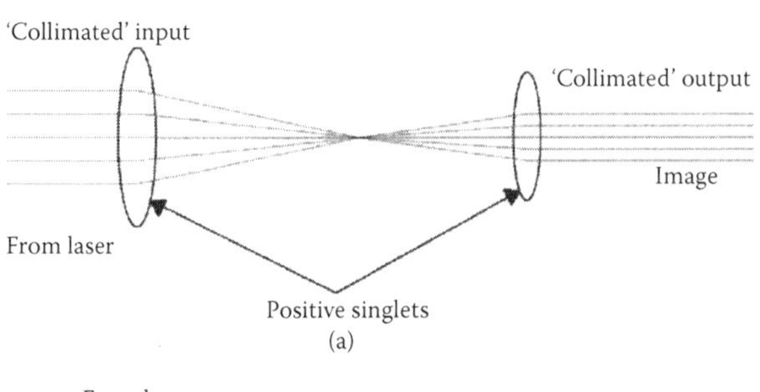

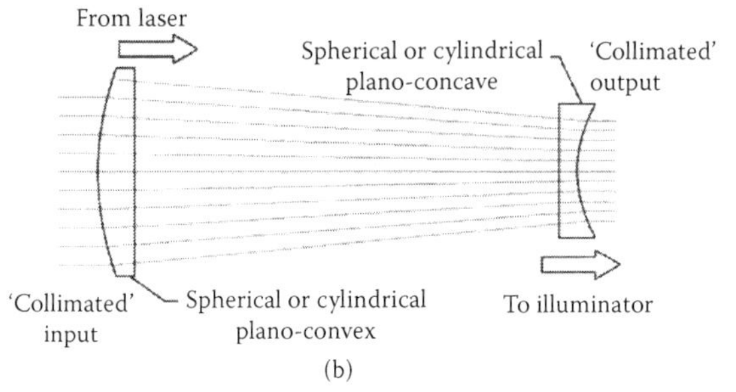

FIGURE 3.8
Telescopes.

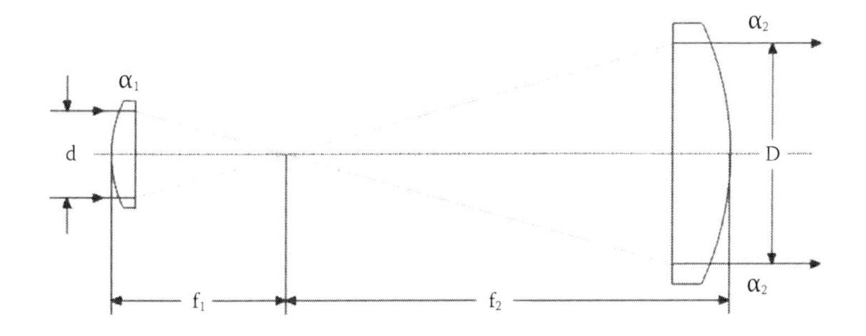

FIGURE 3.9
Keplerian telescope.

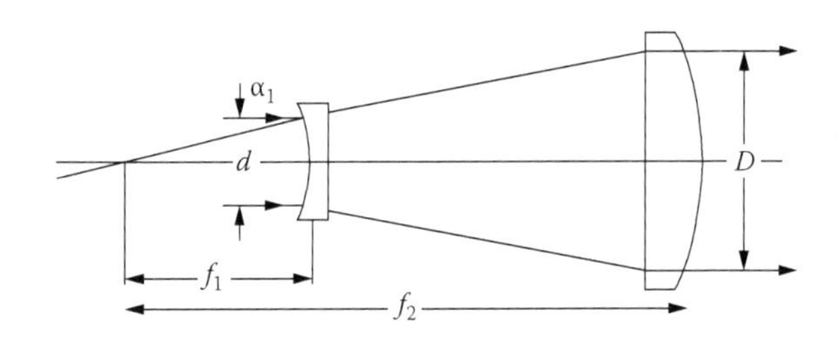

FIGURE 3.10
Galilean telescope.

light). Nonetheless, this technique can be used to get higher throughput for a simple mask or to achieve a higher energy density in a "collimated" beam. Telescopes can be made using spherical optics, which change the beam in all directions, or cylindrical optics, which only compress or expand the beam in one direction.

3.1.5 Beam Profilometry

Profilometry is the analysis of the beam cross section. One type of beam profilometer employs a photoluminescent crystal as the detector (Figure 3.11). A beam splitter deflects a small fraction of the laser beam into the crystal for measurement. The intensity of fluorescent emissions by the crystal is proportional to the intensity of the incident light. Hence, a visible image is created analogous to the spatial intensity profile of the beam. This image is relayed to a CCD (charge-coupled device) camera, where it is digitized, processed, and analyzed by a video image processor. The processed image consists of a false color intensity map or three-dimensional histogram.

Beam profilometry has several applications. First, it allows real-time viewing of the laser output while the system is enclosed if the beam profiler is placed near the laser output. This can help with diagnosis of laser or resonator optics problems and is also helpful during resonator optics alignment after maintenance and for long-term laser performance monitoring. Process monitoring for applications sensitive to beam profile (on-target or mask plane diagnostics) can also be done by viewing the beam profile at the mask plane (in

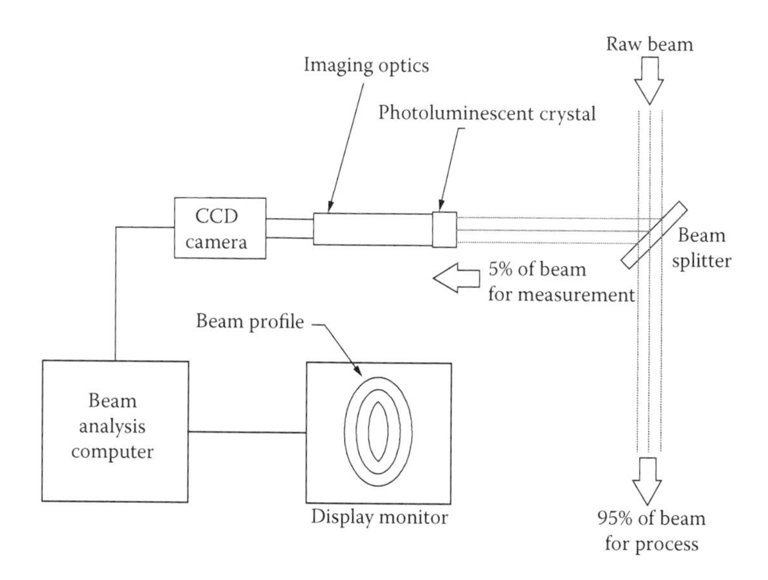

FIGURE 3.11
Beam profilometry.

imaging systems). This permits on-target energy distribution analysis, provides process control information, and verifies proper alignment of the mask.

3.1.6 Homogenizers

The purpose of a beam homogenizer is to break the beam into small sections and recombine them in a pattern that averages the overall energy to produce uniform energy distribution. Some considerations must be taken into account. First, the loss of power density in one part of the raw beam must not affect uniformity of the beam at the mask exposure plane. Significant optical losses are present in any refractive homogenizer configuration, especially as the number of optical elements increases. Collimation is lost: Light typically diverges after the homogenized plane. Homogenizers are expensive and AR coated and the coatings are susceptible to overheating by hotspots within the raw beam. Usually, they are quite inflexible and are dedicated to a particular optical configuration; in other words, there are no really good "general purpose" beam homogenizers. Nonetheless, homogenizers can be very helpful in optimizing laser processing parameters. Various working distances and flat-top sizes are available and, by combining homogenizing with shaping optics, homogenized, shaped beams like line-flat-top, square-flat-top, or spot generators can be realized.

Diffractive optical elements (DOE) can also be used in the optical field of a laser beam and the beam's shape can be controlled and changed according to application needs. The microstructure of the DOE acts like a router for photons, directing their way to propagate through free space. A DOE utilizes a surface with a complex microstructure for its optical function. The microstructured surface relief profile has two or more surface levels. The surface structures are etched in fused silica or other glass types, or embossed in various polymer materials that transmit the laser radiation. Optical losses in diffractive optical elements are low compared to those in refractive optics. Diffractive optics can realize almost the same optical functions as refractive optics, such as lenses, prisms, or spheres, but they are much smaller and lighter. On the negative side, they are more costly and somewhat inflexible because they are generally designed for a specific task rather than for multitasking.

Figure 3.12(a–c) shows three types of refractive homogenizers used with different lasers. A rooftop prism can be used to split a Gaussian beam into two halves and then recombine them such that a flat-top beam is obtained (Figure 3.12a). Cylinder lenses can also be used (Figure 3.12b) to fold the outer wings of a Gaussian beam or even a flat-top beam with "wings" onto the center. Sometimes these wings are filtered simply by using an aperture, which is inexpensive, but adds to optical losses. For the best homogeneity of an excimer beam, a multielement arrangement of cylinder lenses is used (Figure 3.12c), but while the homogeneity is very good, these optics are expensive and inflexible and need to be replaced periodically. Figure 3.13 shows a Gaussian 355 nm laser beam that has been transformed into a square flat

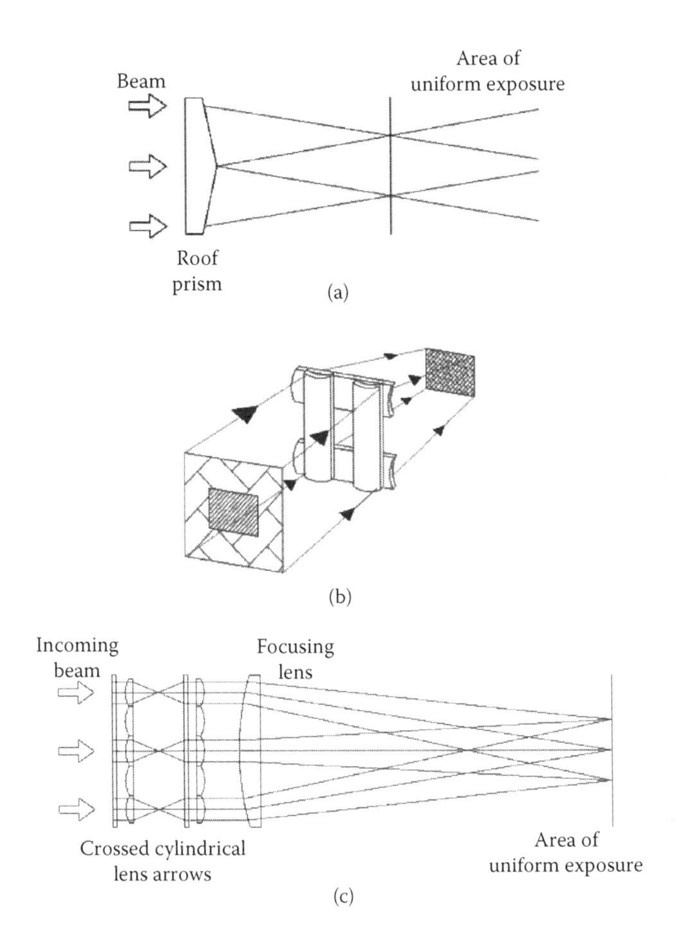

FIGURE 3.12
Homogenizers: (a) rooftop prism; (b) cylinder lens; (c) crossed cylinder.

top using diffractive optics. With a change of diffractive optics, this can also be turned into a round flat top, which is more useful if galvos are used to deliver the beam.

3.1.7 Polarizers

A polarizer is a device that converts a beam of light with undefined or mixed polarization into one with well defined polarization. Linear and circular polarizers are the most common types. Linear polarizers can also be used as beam splitters because they will preferentially pass through light in one orientation with respect to another. Sometimes it may be beneficial to polarize light linearly, but more frequently it is required that light used for micromachining purposes be circularly polarized. In some materials, the polarization makes little difference in the cut quality, but in most cases there is a clear

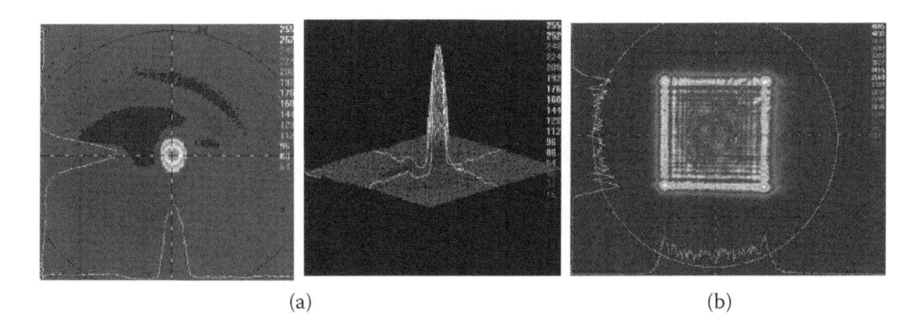

(a) (b)

FIGURE 3.13
(a) Gaussian beam profile; (b) homogenized flat top.

difference in the processing results, especially on metals. This effect can be seen by laser etching a cross and looking carefully at the cut width and quality in both directions.

There are several ways to create circularly polarized light. The cheapest and most common involves placing a quarter-wave plate after a linear polarizer and directing the beam through the linear polarizer. The linearly polarized light leaving the polarizer is transformed into circularly polarized light by the quarter-wave plate.

3.2 Beam Delivery Systems—Imaging and Focusing

The beam delivery system has three main functions: It gets photons from the laser to the work surface, it shapes and conditions the beam along the way, and it protects the operators. Two methods of laser machining are commonly used in industry. Focal point machining is used with solid-state lasers and CO_2 lasers that do not have longitudinal electrodes. Near-field imaging is possible with excimer lasers and TEA–CO_2 lasers because the multimode emissions offer a uniform cross-sectional energy density. A homogenized Gaussian beam or a nearly homogenous part of a Gaussian beam can also be imaged.

3.2.1 Focal Point Machining—Fixed Beam

This is probably the most common method of delivering a laser beam to the work surface. Focal point machining can be done with a fixed-beam delivery system or by using scanning systems such as polygons or galvanometers. For a fixed-beam delivery system, the beam usually enters through a turning mirror and is then directed toward the part, as in Figure 3.14. A fixed-beam delivery will give the smallest attainable spot size (d_{dl}) on target, calculated by the following equation:

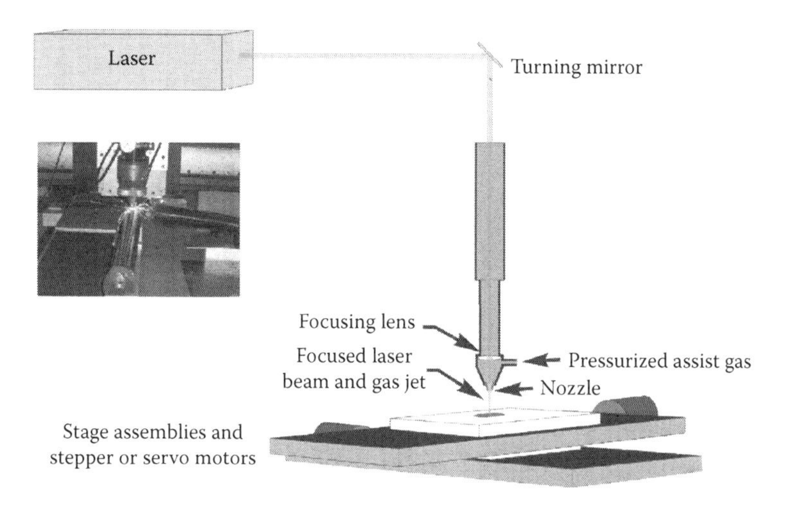

FIGURE 3.14
Fixed-beam delivery.

$$d_{dl} = 1.27\frac{\lambda f M^2}{D} \tag{3.3}$$

where
1.27 = $4/\pi$ is a constant
λ = the laser wavelength
f = the focal length of the lens
M^2 = a measure of the focusability of the laser (provided by the laser manufacturer)
D = the clear aperture of the lens

Slightly underfilling this aperture produces the smallest spot size; significant underfilling will produce a correspondingly larger spot. Also, if the laser beam exactly matches or overfills the input aperture, diffraction effects must be considered and the constant becomes 1.83.

Working with a fixed-beam delivery has some other advantages. First, an assist gas (described in detail later) can be delivered to the part coaxially with the laser beam. This can help cool and clean the part. Also, when used in conjunction with a high-accuracy X–Y table, the best accuracies can be obtained since the accuracy will depend primarily on the X–Y encoders.

3.2.2 Focal Point Machining—Galvanometers

Galvanometer beam delivery systems are used for well collimated beams with low divergence. In such a system, the beam enters the aperture and then encounters two nearly orthogonal turning mirrors mounted on galvanometers that work together to direct the beam through a flat field lens (special cases include f-theta and telecentric lenses) and onto some defined field on target (Figure 3.15). The field size is determined by the choice of laser and optical systems and is limited by focal length, scan angle, and clipping of the beam inside the objective housing. Typically, the usable field size is 0.5–0.7 × the focal length. Clipping happens at smaller scan angles/field sizes with larger mirrors due to a larger beam diameter but also due to a larger separation of the two scan mirrors.

For a simple two-dimensional galvo system, lenses are available from 266 nm to 10.6 μm wavelength; focal lengths of 30 mm to 2 m are commercially available. If no off-the-shelf objective is available or if large, multielement lenses are too costly, a three-dimensional processing approach (discussed in detail later) can be used. A galvanometer-based beam delivery will give a somewhat larger spot size on target than a fixed-beam system and is governed by the following equation:

$$d_{dl} = 1.9\frac{\lambda f M^2}{D} \tag{3.4}$$

The factor d is the Gaussian beam diameter in this case—not the scan head aperture. The factor 1.9 can actually be from 1.5 to 1.9 depending on a few things, but it represents the worst-case scenario. Using 1.9 takes into account aberrations by clipping of the input aperture, aberrations of the objective, and the nonperpendicular angle of incidence. For telecentric lenses, nonperpendicularity does not apply. Aberrations from clipping happen if the beam diameter is larger than one-half of the limiting aperture (scan head

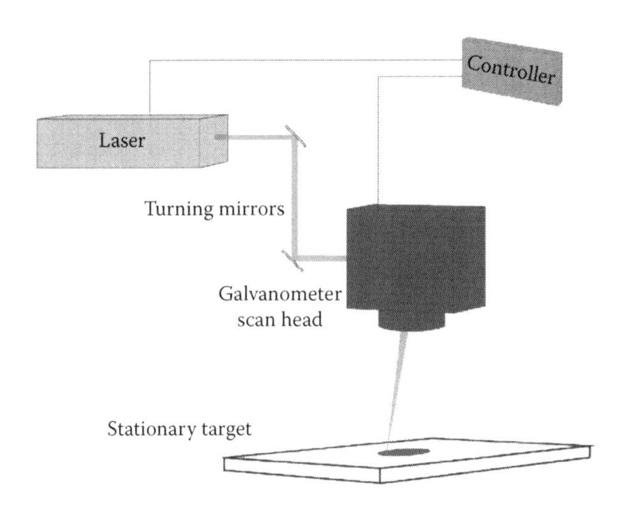

FIGURE 3.15
Two-dimensional galvo beam delivery.

aperture), and aberrations of the objective occur if imaging is not diffraction limited for the beam diameter. In other words, if a small enough beam diameter or a large enough scan head aperture/objective design diameter is used, the spot size is described as in Equation 3.4.

In practice, though, this is sometimes impractical as the systems use larger mirrors and objectives and become slower and more expensive. Typically, some compromises are made in spot size, power losses through clipping, scan head dynamics, and price. Therefore, the beam is usually expanded to from 70% to 100% of the scan head aperture, which gives the factor 1.5–1.9 in Equation 3.4. The big advantage to a galvanometer system is that it is extremely fast compared to table motion. It is not possible to use a coaxial assist gas, so in this case if a cover gas is needed, it must be delivered from the sides or by some other method. There is a practical limit on the field size (regarding spot size, input aperture, aberrations, and price!), but this can be overcome by using galvanometers in conjunction with an X–Y table so that the field can be stepped over the whole table surface if needed. This can be done using a step-and-repeat approach where the same pattern is used for each field, or with different patterns in each field that are "tiled" together to make the whole larger part.

The accuracy of galvanometers is less than when tables are used by themselves and depends on the focal length of the scan lens used. The true accuracy is a rotary function of the galvanometers and the resolution or smallest step size is about 1:65,000 for a 16-bit system, although scanners are available with up to 19-bit resolution. When this is combined with the field size, one can get an idea of the exact step size on target. Resolution alone, though, does not define precision as there are also concerns with dither and drift. It is fair to say that ±10 μm is typical for f = 100 mm scan lens. This equals 1/6,500 scan angle but because some objectives clip the field, 1/5,000 is probably a better number to use. In general, it is difficult to get better than about 10 μm absolute accuracy unless the field is small and active feedback is employed.

For larger fields, a three-axis system can be used. This configuration is shown in Figure 3.16 and employs a moving objective lens plus fixed focusing lens before the scanner mirrors dynamically coupled to the galvanometer mirrors to flatten the field. Because the third axis is slower than the galvanometers, three-dimensional systems can be slower than two-dimensional systems, but only if the third axis is limiting the speed. For instance, marking concentric circles or making rapid movements within a very small area (within the depth of focus of the optic) with a three-dimensional system is not slower, but larger movements over the field are slower. Of course, the longer distance to the target also produces larger spot sizes unless larger apertures are used to compensate for the extra distance. CO_2 systems used for cutting web material like labels and stickers use this set configuration.

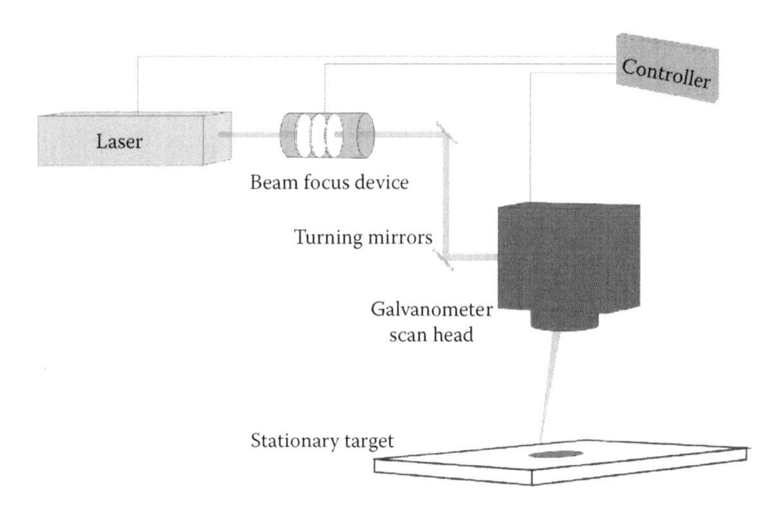

FIGURE 3.16
Three-dimensional galvo scanning beam delivery.

3.2.3 Near-Field Imaging

Near-field imaging involves use of a mask to project a pattern of light onto a part (Figure 3.17). The features of the mask are etched into the target material at a magnification determined by the relative positioning of the optical elements, and the optical system is described by the thin lens equation (Section 3.2.5). Near-field imaging is used when the beam is multimode and reasonably homogeneous—at least in some areas of the beam. Fluence and spatial distribution are controlled by optical magnification. The setup is fairly simple and high tolerances are achievable since imperfections in the mask are demagnified. Theoretically, imaging resolution is on the order of emission wavelength but, in reality, practical imaging resolution is dependent on optics. Imaging can be done using spherical or cylindrical optics. Table 3.1 shows some advantages and disadvantages of this type of imaging.

3.2.4 Masks

Imaging systems require masks. They can be very simple, such as a round hole in the center of a piece of metal, or extremely complex with unsupported features. Masks must be made to fit within the homogeneous portion of a laser beam in order to achieve uniform illumination on target. Simple masks with supported features can be made on a laser using thin stainless steel or copper. Chemical etch can also be used to make masks in thin metals. The attainable resolution of these masks is not very high, but many can be etched on one sheet with varying dimensions if a variety of masks is needed. Metal should be as thin as possible in order to avoid or at least minimize diffraction effects. Since any defects present in the mask are demagnified in accordance

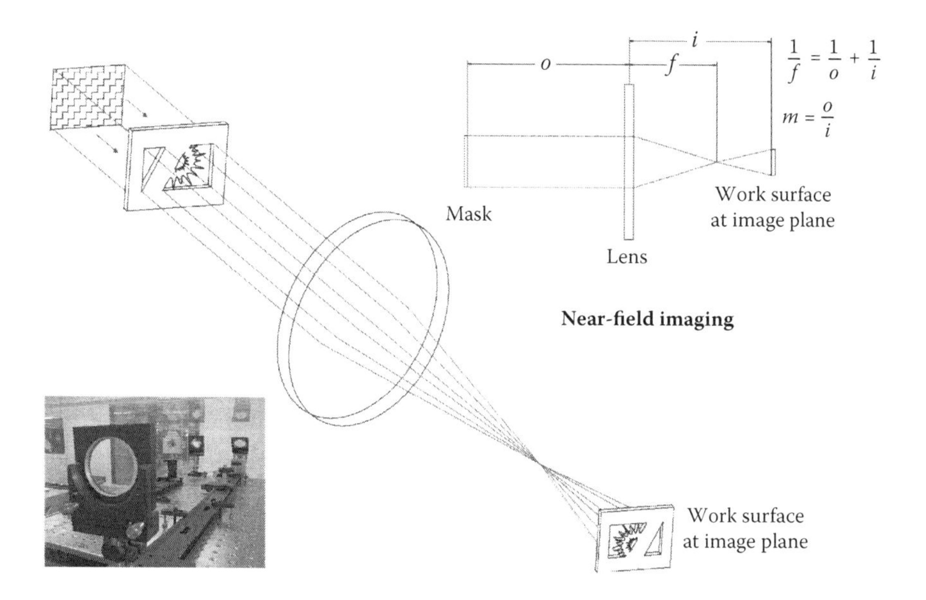

FIGURE 3.17
Near-field imaging.

TABLE 3.1

Advantages and Disadvantages of Near-Field Imaging

Advantages	Disadvantages
Very flexible for a wide range of shapes	Mask must fit into usable portion of the beam
Fairly simple optical setup	Focus is critical to feature quality
High tolerances can be met	Spherical aberration can distort image shape
Choice of different masks for different applications (metal, chrome on quartz, dielectric)	Energy density nonuniformity across the mask is duplicated on the part
Very wide range of demagnification possible	Work area on process material limited by demagnification

with the optical system demagnification, fairly high resolution images can be obtained on target with even simple masks.

When unsupported features are present or when the features are very small (especially in a low demagnification optical system), it is not possible to use a simple metal mask. Unsupported features require that the mask be supported on a transparent substrate. The mask in this case is a thin layer of either a robust metal or a high reflective dielectric coating. These masks are much more expensive than a simple metal mask, but much higher resolution features are possible and diffraction from the very thin edges is minimal. Holographic and diffractive optical elements can also be used as masks.

3.2.5 Thin Lens Equation and Demagnification

The relationship between the mask, imaging lens, and image on the target material is described by the *thin lens equation:*

$$\frac{1}{o} + \frac{1}{i} = \frac{1}{f} \tag{3.5}$$

where
o = object distance
i = image distance
f = focal length (normally expressed in millimeters)

In optics, the term "magnification" is normally used to describe the ratio of the image size to the size of the object. Magnification implies that the image size is greater than the size of the object. The term "demagnification" implies that the image is smaller than the object and is described by the following equation:

$$d = \frac{o}{i} \tag{3.6}$$

where d is the demagnification.

We can relate the object and image distances to the demagnification by substituting the demagnification equation into the thin lens equation:

$$o = (d + 1)f \tag{3.7}$$

$$i = \frac{(d+1)}{d} f \tag{3.8}$$

3.2.6 Beam Compression

A cylindrical lens is normally used for scanning a large area with a relatively low fluence and for planarization. If a cylindrical optic is used to compress the beam, then the dimension of the beam changes only in one direction. The fluence after cylindrical compression, ρ_1, is given by

$$\rho_1 = \rho_0 d(1 - L_f) \tag{3.9}$$

where
ρ_0 is the initial fluence before the lens
d is the demagnification factor
L_f is the percentage loss through the optic

Uncoated optic losses typically run about 5% per optical element, so it is important to minimize the number of optical elements in the design of the optical system.

If a spherical optic is used to compress the beam, then the dimension of the beam changes in both directions. In this case, the fluence after spherical compression ρ_1 is given by

$$\rho_1 = \rho_0 d^2(1 - L_f) \tag{3.10}$$

where
ρ_0 is the initial fluence
d is the demagnification factor
L_f is the percentage loss through the optic

Size constraints placed upon the optical systems limit the achievable demagnification factor and fluence for a given lens focal length.

Some factors limit the usability of near-field imaging. If the optical path is very long, the available space may be insufficient in length to accommodate a specific demagnification for a given lens. Laser energy can be insufficient to photo-ablate certain materials for a specific demagnification. One axis of a noncircular beam may be too small to accommodate a mask required for a specific demagnification or image size. The required image and mask size can be small enough to create undesirable diffraction interference at the image plane. This condition occurs in metal masks when feature sizes are small (it depends on optical design, but for a typical machining system this means features < 0.004 in.) and is a particularly serious limitation if the image is a repeated pattern of holes or shapes.

3.2.7 Beam Utilization Factor

When a mask imaging system is used, much of the beam is lost because it hits the mask and gets reflected, so not all of the available photons are used for processing. Another consequence is that the mask may heat up, so cooling of the mask may be necessary to avoid distortions. It is best if the optical setup has a high beam utilization factor (BUF). Figure 3.18 shows a diagram of a raw laser beam and the usable fraction of that beam (UFB). The mask is then put into the homogeneous area of the beam and only the light that strikes the mask opening(s) is propagated to the target.

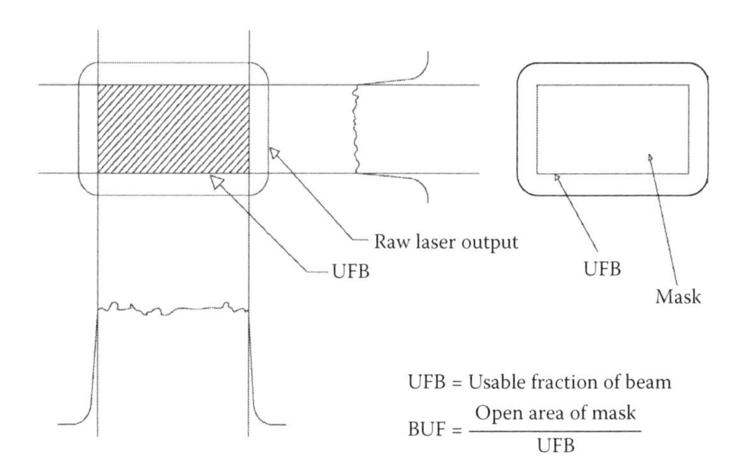

FIGURE 3.18
Beam utilization factor.

For example, assume that the UFB is about 1 × 2 cm and the mask is a circle with a diameter of 1 cm. If an area ratio between the mask opening and the UFB is taken, we find that the beam utilization is about 39%, so over 60% of the photons are lost at the mask, decreasing the efficiency of the system. Clever beam shaping or splitting and/or using holographic or diffractive optics can increase the BUF. Optimum beam utilization is essential to quality part manufacturing at affordable costs. The definition of UFB is application dependent as one process may require tighter control than another.

Other losses in the optical train must be addressed. Refractive optical materials absorb a small fraction of the beam. This absorption depends on optic thickness. Reflection from surfaces of refractive optics contributes to losses. Absorption losses can be minimized by proper choice of optical materials and high-quality optics workmanship.

3.2.8 Beam Optimizing Considerations

Optimum beam utilization is essential to quality part manufacturing at affordable costs. Techniques for optimizing BUF include parallel processing schemes in which a laser beam illuminates multiple imaging systems to process many parts simultaneously. The size of the UFB is highly dependent on resonator optical alignment, and the definition of UFB is application dependent. One process may require tighter fluence control than another. Refractive optical materials absorb a small fraction of the beam, but with the proper choice of high-quality materials, refractive optics can approach 99% transmission. Absorption depends on optic thickness.

In addition, reflection from surfaces of refractive optics contributes to losses. Reflective optical components have losses of 1%–2%. Absorption losses can be

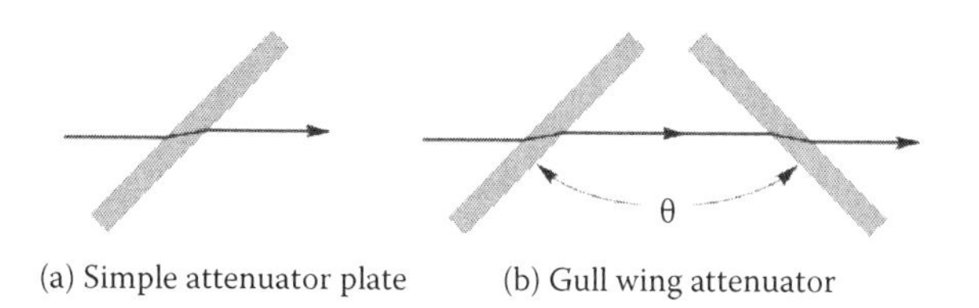

FIGURE 3.19
Simple and gull wing attenuators.

minimized by proper choice of optical materials and high-quality optics workmanship. Reflective losses with refractive elements can be minimized with angle- and wavelength-dependent high-reflection dielectric coatings. Losses are best controlled by limiting the number of optical elements, and refractive optics should be as thin as possible. More elements are acceptable if the process requires high precision and beam utilization is a secondary consideration.

Automation of beam delivery components provides a powerful method of speeding part processing with laser systems. The following list describes just a few of the automation alternatives involving beam delivery optical setups:

Autofocus. The lens is focused automatically by a stepper motor controlled by the system computer at the touch of a key or through a process program.

Automagnification. The mask and lens are slewed together according to algorithms derived from the thin lens equation.

Autoattenuation. A simple attenuator can be made using a single partially transmitting plate and rotating it in the beam. Note that, because of refraction, there will be beam "walk off" as the plate is turned, so it is only useful in a static mode. When dynamic attenuation is needed, a pair of attenuator plates (Figure 3.19) is used in a "gull wing" fashion so that refraction through one plate is offset by refraction through the second plate and the beam does not walk off target.

Rotary or linear mask control. A rotary mask or linear mask assembly is actuated by a computer-controlled stepper motor or pneumatic device. Mask alignment permits high-precision overlays, such as counterbored holes. Mask aligners with letter stencils can produce custom serialization marks under computer control.

Galvanometer beam steering. As discussed previously.

3.2.9 Coordinated Opposing Motion Imaging

In coordinated opposing motion imaging (shown in Figure 3.20), both the mask and part are mounted to computer-controlled X–Y stages. During

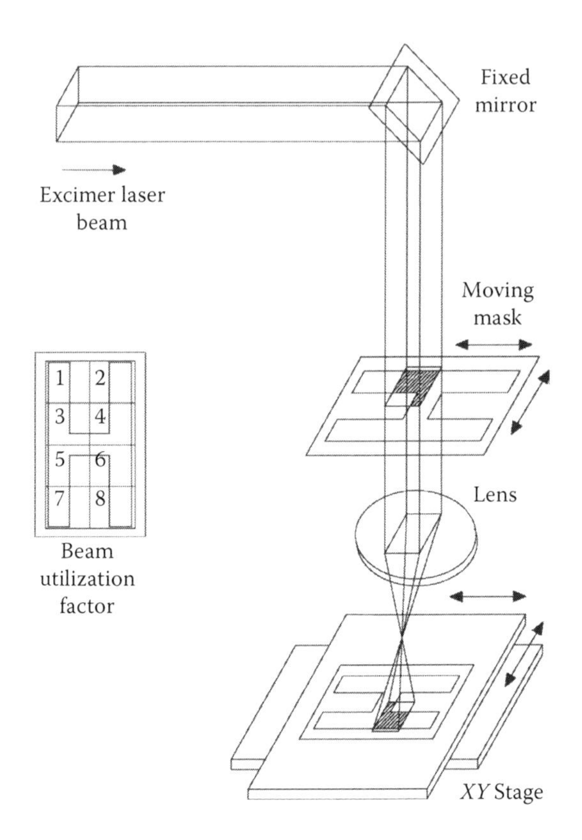

FIGURE 3.20
Coordinated opposing motion imaging.

processing, the mask and stages perform interpolated moves in opposing directions; the magnitude of mask movement is smaller by a factor equal to the imaging system demagnification. This opposing motion causes the laser image to track the position of the moving part precisely, remaining at the same position relative to the part as different areas of the mask are exposed. The lens is always used in the paraxial, on-axis condition. Scanning can also be done in only one dimension if necessary.

3.2.10 Direct-Write Machining

Direct-write machining provides a useful technique of generating large cutout features and performing high-volume hole drilling in materials where fluence requirements limit spot size. The required features can be drawn in computer-assisted design (CAD) and then directly translated to motion control code utilizing a CAD/CAM (computer-aided manufacturing) programming interface (Figure 3.21). Specific functions possible using this technique include the following:

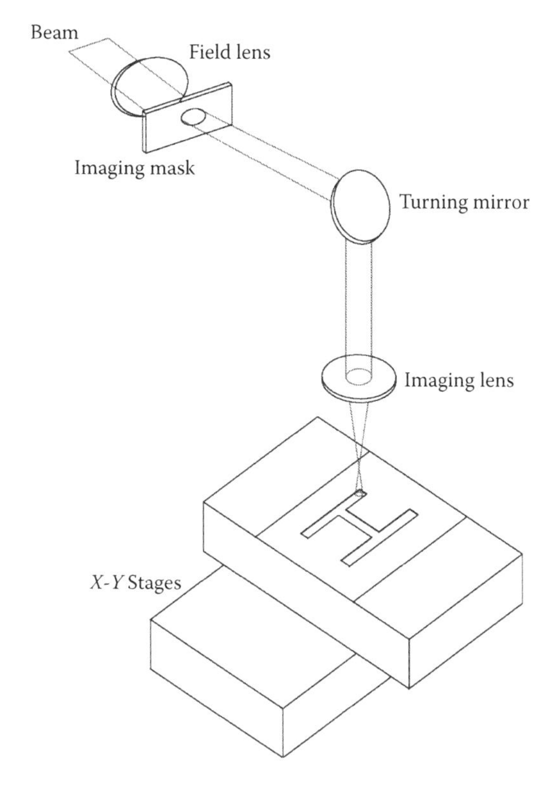

FIGURE 3.21
Direct-write machining.

- Features on the CAD drawing can exist on different layers that correspond to automatic mask changes or changes in laser pulse spacing within a single process program.
- Points can be placed in the CAD drawing to trigger step-and-repeat drilling operations.
- Drill marker points can be drawn on different layers within the CAD file to control drilling depth.

3.2.11 Contact Mask Processing

Contact mask processing is a technique by which fluence on target is controlled by simple beam shaping optics and feature shape is determined by a blocker mask in contact with the workpiece (Figure 3.22). The exposure of the contact mask may be performed with the part stationary or scanned under the beam. If scanning is used, laser firing must be interpolated with the table feed rate to ensure uniform exposure. Contact mask scanning allows very large areas of material to be processed.

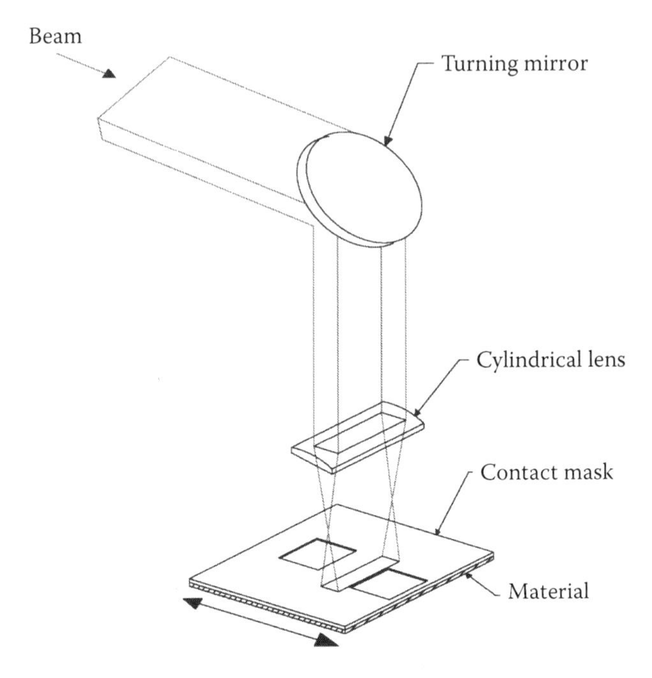

FIGURE 3.22
Contact mask processing.

There are some considerations in contact mask processing. The blocker mask material must be selected so that the beam does not damage the mask as it ablates the material. Some choices include copper, stainless steel, aluminum, and molybdenum. If the mask is not expendable, select the minimum fluence required to minimize the damage. Periodic cleaning of the mask edges is required for maintaining sidewall quality. Feature resolution is limited by the limits of the size of the feature that can be put into the mask. Features down to 25 μm have been achieved. A conformal mask can be laminated to the workpiece as an integral component of the final product. This technique is described in detail later, related to formation in microelectronics packaging.

3.2.12 Multiple Laser Beams

Beam dividing can be used to increase the BUF as explained before. Figure 3.23 shows another technique: the use of multiple lasers on a single production line. Seven lasers and separate beam delivery lines are used with each beamlet shuttered to produce continuous marks on a moving production line. This technique has been used with up to 80 lasers firing simultaneously on a single production line.

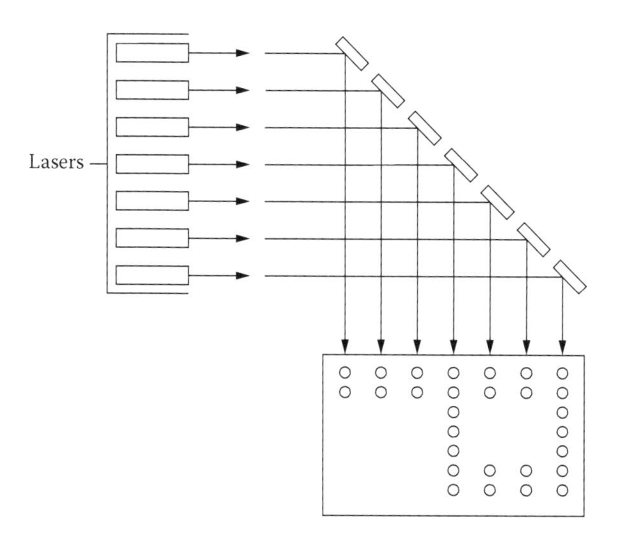

FIGURE 3.23
Online marking with seven beamlets.

3.3 Steps to an Effective Optical Setup

Determine which type of laser is best.

Are machining quality and tolerances tight enough to warrant UV laser use or can an IR laser do the job?

Is the target material thin enough for UV laser machining, or is an IR wavelength required?

What is the best wavelength? Should more than one laser be used? Some products may require the use of multiple lasers, each dedicated to a specific task.

How much pulse energy and power are required?

If excimer laser imaging is chosen, determine the wavelength and energy required. Is 193 nm, 248 nm, or 308 nm best absorbed? What is the ablation threshold of the material? Experiment to determine optimum fluence and wavelength, if possible.

Determine required exposure area.

What are the required feature sizes?

Can a feature be broken into smaller portions?

How long will it take to accomplish the job?

Determine resolution and optical setup type.

- Contact mask or imaging system (long working distance or microscope system)?
- Which objective is best?
- Select an objective that minimizes optical losses.

Determine pulse energy required.

- Required pulse energy on target = required fluence × feature area.
- Do not forget to factor in optical losses.

Determine optical parameters.

- Select demagnification.
- Maximize BUF; use beam shaping/splitting as required.

Determine optical path; use readily available optics.

- Consider size of the beam delivery system and support structure.
- How critical is beam uniformity? Homogenizer? Consider losses.
- Do not forget to factor in taper effect.

Select size of laser and illumination scheme.

- Laser energy required = required pulse energy on target × BUF.

Select lowest pulse energy laser to do the job.

- What if required pulse energy is greater than maximum output of available lasers?

Reconsider optics scheme to reduce losses.

Consider a scanned illumination technique.

4

Light–Material Interaction

Light or photons interact with materials in very specific ways. This interaction depends upon the properties of both the material and the light and how this light is delivered. Important properties include wavelength, intensity, material absorption (at a specific wavelength), material thickness, and optical configuration. This chapter discusses material–light interactions in detail.

4.1 Photoablation and Material Interaction with UV Light

Figure 4.1 shows a graph of etch rate versus fluence (on-target energy density in joules per square centimeter). The shape of this graph is typical for any material and laser combination, and generating such a graph is a very useful first step when dealing with unknown materials. At very low energy density, there is no material removal until a point is reached where material removal starts to occur; this is called the *ablation threshold.* Below this threshold, the photons are still striking the material and, assuming that they are not reflected, the photon energy gets absorbed by the material, but is not used for material removal. The energy contained in the photons is therefore imparted as heat into the material or potentially changes the surface structure.

The general shape of this curve is similar for all lasers and materials, but the ablation threshold point and the curve plateau will change depending on absorption. For instance, let us assume the curve shown is for a material with absorption α_1. If another material has an absorption α_2 and $\alpha_1 > \alpha_2$, then the ablation threshold for α_2 will be higher and the plateau or maximum etch rate will also be higher.

For instance, with UV light, a photochemical color change can occur in polymers and ceramics at low fluences. In this case, UV light alters the surface molecular structure, leading to a change of light-absorbing properties and a color change results. Once the energy density is sufficient to overcome the ablation threshold, the per-pulse removal rate climbs until a point is reached where the ablation rate per pulse approaches a constant. On this plateau, there is a good working area since working at lower fluences (on the up curve) will cause fluctuations in the material removal rate with fluctuations

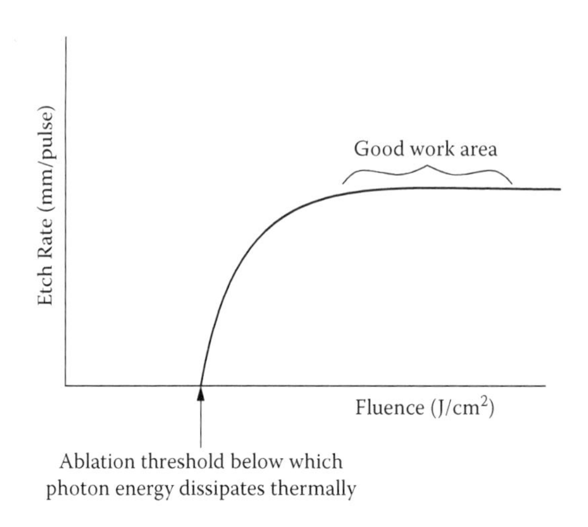

FIGURE 4.1
Etch rate versus fluence.

in the laser output; working far out on the plateau gives no obvious benefit in removal rate. Each material has its own ablation threshold, below which photoablation does not occur. Pulsed lasers remove a certain amount of material per pulse, and this process is repeated pulse by pulse, as in Figure 4.2. Ablation by-products and some excess heat are carried away by expansion of the plasma plume.

4.2 Thermal Effects

One of the most important factors in laser micromachining applications is the edge quality. In a perfect world, for a cut or drilled hole, one would want straight, smooth side walls, little to no taper, and no change of characteristics in the surrounding material. The latter is almost always caused by excessive heat being imparted in a time short enough such that the heat cannot be easily dissipated. Thermodynamics tells us that work, potential energy, kinetic energy, and heat are all manifestations of the same thing; that is, they all have the same physical units, and energy can be transferred back and forth between them (although not 100%, as we also know by considering the entropy term).

Each photon possesses some energy, *E*, that can be transferred to matter. How this energy transfer occurs is important. If the energy of every photon can be used to dissociate chemical bonds efficiently, then no residual heat will be left to cause problems in the material. If, however, the photons do not do work, but instead generate heat, the surrounding material may suffer damage. Therefore, in order to minimize HAZ (heat-affected zone), efficient

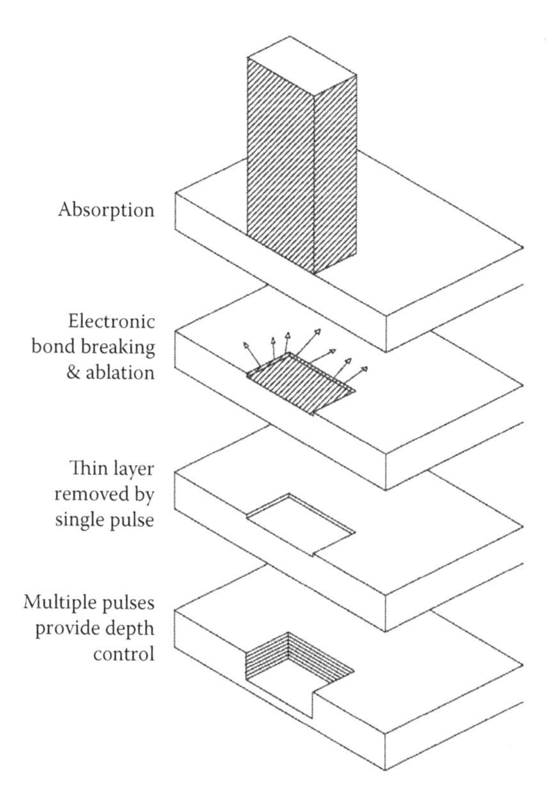

FIGURE 4.2
Photoablation process by exposure to UV light. A single ablation layer is typically <1.0 μm thick, due to the short UV wavelength.

processes must be used in the first place and then consideration given to how to minimize any residual heat affects. So, what factors contribute to HAZ? These factors can only come from two sources: the laser and the material. Each factor is discussed next.

The laser

- Factors that may contribute to heat generation on target that are laser related are wavelength, pulse length, energy per pulse, repetition rate, and beam shape. The laser wavelength determines the photon energy; the shorter the wavelength, the higher the photon energy. Infrared (IR) photons interact with vibration–rotation modes of the material while ultraviolet (UV) photons can break chemical bonds directly. Therefore, as a general rule, one goes to shorter wavelength in order to minimize heat effects. For the sake of argument, assume that a UV laser is used so that the first-order effects on the chosen material are not thermal.

- The next consideration is pulse energy. Per-pulse laser energy must be high enough to initiate ablation, but not so high that all the photon energy cannot be used efficiently. This condition is met when operating in the "good work zone" on the ablation curve (Figure 4.1). The pulse length is important in that high peak power intensity results from short pulse length. The shorter the pulse length, the more likely it is that the photon energy will be used to remove material efficiently rather than to impart heat into the surrounding material. It should be noted that for very short pulse lengths (sometimes less than 1 ns and surely less than 100 fs), wavelength dependence is also minimized. This may be the result of higher order nonlinear effects in the material. The pulse repetition rate is also important in that ejected material needs some time to escape, and if the pulses are too closely spaced, this cannot occur. High repetition rates (megaHertz) normally are not useful and are frequently harmful in micromachining applications, as the beam cannot be moved fast enough. Optimizing the laser repetition rate is also related to pulse length because the shorter the pulse length, the faster the plasma will escape.
- Finally, there is beam shape. What is desired on target is a flat-top beam, but this is generally not the case. In fact, most lasers have a Gaussian beam profile, as in Figure 4.3. The laser spot size is generally meant to mean the FWHM (full width, half max) of the Gaussian profile. This profile is the typical standard distribution in mathematics. The actual entrance diameter on target of a percussion drilled hole will be close to this value, but not identical because it depends on how much energy density is required to remove the material. Therefore, the hole on target may be smaller (d') if more energy density is required to do the machining and it may be bigger (d'') if less energy density is needed. In the case of

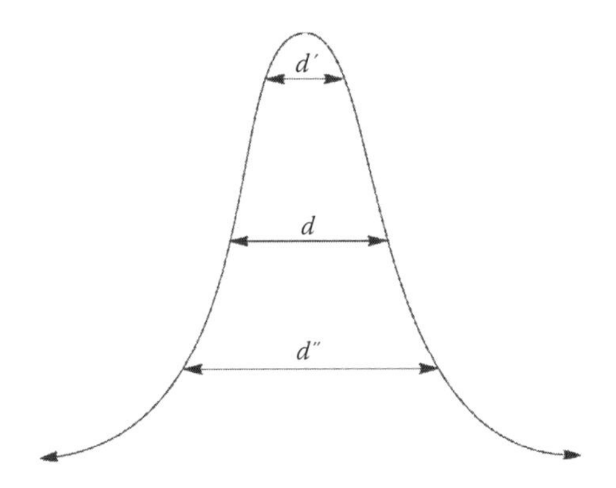

FIGURE 4.3
Gaussian laser beam profile.

d', especially, this means that all the photons in the "wings" of the Gaussian beam are imparting their energy into the material in the form of heat rather than performing the work to remove material.

The material

- The first and most important attribute of the material is that it must absorb the incident photons; the higher the absorption, the better the energy transfer into material removal rather than into thermal dissipation. There is really no way to change this as it is a fundamental property of the material with respect to the wavelength used, but adding adulterants with strong absorption can help.
- The bonding structure is also important. For instance, polymers with π-electron clouds strongly absorb UV light. However, metallic bonds do not necessarily absorb the UV photons as efficiently, so thermal effects are seen on metals even at short wavelengths.
- Another important property is the thermal conductivity. It is better not to generate heat in the first place, but if heat is generated, a material with strong thermal conductivity will show better edge quality and less HAZ.
- Things get even more interesting with a nonhomogeneous material such as FR4 (flame-resistant class 4)—glass fibers embedded in resin. Ablating glass fibers requires a very high pulse energy laser, much higher than is needed for the surrounding epoxy matrix. This can impart undesired thermal side effects into the surrounding material.

So, in principle, if HAZ is a real concern, it is best to choose conditions, lasers, and material that are more conducive to this processing. If this cannot be done, then some additional processes may need to be used, such as assist gases.

HAZ is a very important consideration, especially when laser processing small and/or high-value parts. When processing quality concerns outweigh other factors (like fiscal!), there are ways to minimize HAZ. The basic rules are shorter wavelength, shorter pulse duration, high per-pulse energy, uniform illumination, and proper match of laser to material.

4.3 Taper

When lasers are used for materials processing, taper is inherent in the final result. This taper is usually oriented as shown in Figure 4.4, where the laser entry side is larger than the laser exit side (or bottom, if the feature is "blind"). For simplicity, consider the case of laser drilling round holes. When a round

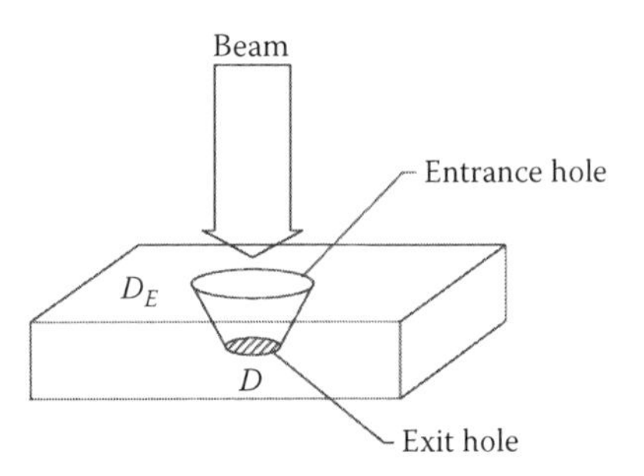

FIGURE 4.4
Taper.

hole of a certain diameter is desired, this diameter is usually specified at the exit side, labeled "*D*" on Figure 4.4, and is determined by the taper angle. The entrance hole diameter, shown as "D_E," is determined by the optical setup using mask imaging, percussion drilling, trepanning, or some other rastering technique. This taper angle is usually between 2° and 10° and is influenced by a number of factors that will be considered next.

The laser beam can be delivered to the work surface in a number of different fashions. Figure 4.5 shows three of the most common hole-drilling methods. All of these methods can be used with either fixed-beam or galvanometer steered-beam delivery systems, although in practice the first two are probably used more in fixed-beam systems, while the third is used primarily with galvanometers. Single-shot drilling gives a hole of approximately the diameter of the incoming beam and is the fastest method of drilling holes. The laser energy and material absorption must be such that the material can be etched with a single shot. Taper tends to be somewhat more in this case as there are no subsequent pulses to the initial pulse to "clean" the hole. Percussion drilling uses multiple pulses to etch through the material. This process is slower, but it also produces a rounder hole with less taper. Finally, the laser beam can also be trepanned—very rapidly when using a galvanometer—to produce a hole with the best circularity and the lowest taper.

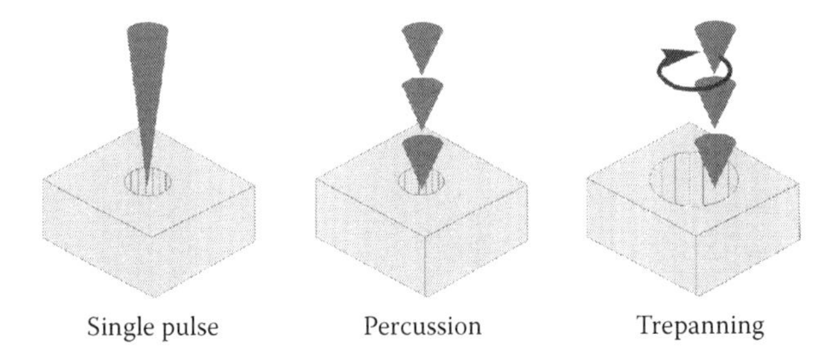

FIGURE 4.5
Laser drilling techniques.

In addition to the method of beam delivery, other factors affect taper. These include peak laser power density (defined as energy per pulse divided by pulse width multiplied by the area on target), pulse repetition rate, number of pulses, assist gas pressure (if used), focus position, and aspect ratio. As a general statement for any kind of machining laser process, the higher the peak power, the better the results. Higher peak powers give correspondingly lower taper, to a point; "zero taper" is very difficult to achieve without using special optics. Therefore, by correlation, higher energy per pulse and shorter pulse widths also will tend to give less taper. Since pulse repetition rate is the number of laser pulses delivered to the target per second, this has no meaning in single-shot delivery, but there can be an effect with the other two delivery methods.

In principle though, as long the repetition rate remains low enough so that the beam can be moved quickly relative to the pulse frequency for heat dissipation, the effect on taper should be minimal. The number of pulses delivered in a percussion drilling mode has little to no effect on entrance hole *D*, but it has a big effect on taper—more pulses giving less taper, again within some boundary conditions.

Focus position is also crucial. For most processes, the focus is at the top surface of the material. In some processes, however, it is more beneficial to focus on the bottom or in the middle; in fact, sometimes a dynamic focus is used (for instance, to go through a thick piece of material). The hole shape in Figure 4.4 is typical of a hole drilled by focusing on the top surface.

Next, consider the aspect ratio. This is defined as the material thickness (or hole depth if blind) divided by the hole diameter. Lasers can be used to make very high aspect ratio features; 10:1 or even 20:1 is common and straightforward, and aspect ratios up to 100:1 can be achieved under the right conditions. This is where lasers differ greatly from chemical etching, where a 1:1 aspect ratio is about as good as can be obtained and even then there is significant taper or undercutting. As the aspect ratio increases, in general, so does taper, so smaller holes and/or thicker material give greater taper. Finally, holes drilled in different materials will show different taper characteristics under the same laser conditions. For instance, materials like polymers with a lower ablation threshold will show less taper than drilled stainless steel using the same laser energy settings because the ablation threshold for stainless steel is much higher than that for plastics.

In addition to a chemical etch, consider some other machining methods and how their inherent taper compares to lasers. One of the biggest drawbacks to abrasive water jet cutting is the taper it creates: about twice that of a laser processed part. Taper can be controlled by slowing down the process, but only to a certain point and at the cost of time and money! Therefore, in practice the part is only slowed down in areas where taper is important. Also consider EDM (electrode discharge machining) or mechanical drilling; both of these processes give essentially taper-free holes, but both also have drawbacks (as discussed in Chapter 1).

Finally, it should be understood that taper is sometimes desirable. Many parts made by laser processing—especially in areas like drug delivery for disposable medical devices and microvias for printed circuit boards—benefit from taper. In the case of fluids, constriction in the flow direction is desirable. In the case of interconnects, the taper helps to get cleaners and fillers into the holes to form the electrical connect. In fact, optical methods are available that will even introduce negative taper (entrance hole smaller than exit hole) using rotating prisms or offset beam plates. In summary, hole taper (and by extension feature taper) is inherent in laser processing, but by controlling the parameters that affect taper, it can be controlled and used to advantage.

4.4 Fluence

The term *fluence* describes the amount of energy per area deposited on the target surface by a single laser pulse. Fluence should not be confused with pulse energy, because the amount of fluence that strikes the target depends on transmission losses in beam delivery components as well as spatial compression of the beam. Fluence is important in laser machining because specific materials require minimum fluence levels to produce material removal. Fluence is defined by Equation 4.1, where ρ is the fluence (joules per square centimeter), E is the energy measure on-target in joules, and A is the area of the image (square centimeters). Fluence can be approximated for focal point applications as well, where A is the area cross section of the beam at the focal point:

$$\rho = \frac{E}{A} \tag{4.1}$$

For any fluence ρ and area a, the product $\rho \times a$ will be a constant, assuming no loss of photons; for example, compressing a laser beam to a smaller area results in a correspondingly higher fluence. This is described by

$$\rho_L a_L = \rho_T a_T \tag{4.2}$$

The left term is the energy density × area product at the laser (L) and the second is at the target (T). For example, for a homogeneous excimer laser beam of dimensions 1 × 2 cm and a fluence of 100 mJ/cm^2, assuming that 5 J/cm^2 is desired on target, the resulting available processing area is described by

$$a_T = \frac{\rho_L a_L}{\rho_T} = \frac{(0.1\text{J}/\text{cm}^2)(2\text{cm}^2)}{5.0\text{J}/\text{cm}^2} = 0.04\text{cm}^2 \tag{4.3}$$

5

System Integration

After all is said and done, a laser is just a glorified lightbulb—a device that emits light. It is, by itself, not very useful for anything until one adds the other ancillary equipment that makes the laser into a machine tool. Laser micromachining is still machining, but instead of using a drill bit, router bit, or saw blade, light is used. This section discusses the other components needed to complete a laser micromachining workstation.

5.1 Processing System Considerations

1. The laser. Will the system be on- or off-line? While off-line systems may share data with other computers, they are otherwise self-contained. Online systems are interfaced to some sort of production line, perhaps incorporating many different laser and nonlaser stations; physical compatibility as well as software compatibility and "handshaking" is necessary. Will the system be located in a clean room? Alternatively, will the environment be uncontrolled? What are the available utilities—electrical, cooling, venting, gases, CDA (clean, dry air)?
2. Beam delivery. Once the laser is chosen, the proper beam delivery must be used in order to maximize utility. Fixed beam? Galvanometers? Automated? Will assist gas be used and, if so, what kind? Multiple wavelengths? Use the information presented in Chapter 3 to decide.
3. Motion control. Are motorized stages used? X–Y? Z? Rotary stages (both lathe type and other orientations can be considered)? What kinds of stages? Travel? Resolution? Galvanometers are also part of the motion control as they interface with the other components.
4. Part handling. Is it a manual load and unload system? If not, will there be robotics, conveyors, roll to roll? Part pallets? Other tooling? Any other considerations? Will a vacuum chuck be needed and, if so, what size and type?

5. Vision. Is a camera needed? If so, is it simply to look at something or is it to be used to align accurately? Single or multiple cameras? Simple or automated vision systems? Color or black and white?
6. Computer and software. Will functions be handled with a computer or programmable logic controller (PLC) or both? What kind? What are the programming language and operating system? Software is key to ease of use of laser tools. What kind of monitor will be used: size, touch screen, etc.?
7. Safety. In all cases, safety must be attended to. Most laser tools are built in a Class I configuration, but some are Class IV. Safety includes considering stray laser light, gases, optics, mechanics, electricity, effluent, etc.

5.2 General Requirements

Lasers have been described in detail in Chapter 2. Determining the laser or lasers is the first step. Most laser machining systems have one laser and are dedicated to processing with a particular wavelength of light. Some systems have multiple lasers (one solar company has incorporated over 80 lasers into one system) or use one laser with multiple wavelengths. In all cases, physical space as well as cooling and electrical requirements must be addressed.

The next consideration is location. It is highly recommended that a laser micromachining tool be located in at least a temperature- and humidity-controlled environment. It is further recommended that a class 100,000 clean area be used, especially for high-precision laser tools, as it keeps the optics and work area clean. The controlled environment helps to keep the lasers running well. Although most commercial lasers are pretty forgiving, some require temperature stabilization to within a few degrees, and humidity is never good for laser optics. If the tool is to be located in a clean room with a cleaner rating—for instance, Class 100—it may require a different choice of materials or vendors to meet compatibility requirements.

A clean room has a *controlled* level of contamination that is specified by the number of particles per cubic meter at a specified particle size. To give perspective, the ambient air outside in a typical urban environment contains 35,000,000 particles per cubic meter in the size range of 0.5 μm in diameter, corresponding to an ISO 9 clean room; however, an ISO 1 clean room allows no particles in that size range and only 12 particles per cubic meter of 0.3 μm and smaller. A class 100,000 clean room has 100,000 particles > 0.5 $\mu m/m^3$ (ISO 8) and a class 100 clean room has 100 particles > 0.5 $\mu m/m^3$ (ISO 5). The more the environment is controlled, the more difficult it is for operators to work (gowning, entering, cleaning, etc.), but also the better the environment is for laser tools as they work best in a controlled environment. This space is premium, though, so sometimes portions of the laser tool (laser, chiller) are outside the actual clean room in a utility chase area. If the environment is *not* controlled, keeping the tools running consistently (with fluctuations in temperature and humidity) and achieving the finest feature resolution may be a challenge.

If the system is to be used "online," at a minimum, physical and software handshaking is required. Parts must be passed from station to station using conveyors, robots, or some other means. Handshaking software is dependent on the hand-off used. For example, consider the case of a conveyor. Normally, sensors detect the presence of a part in the load position. Another sensor detects parts passing the unload position. A new part is loaded into the laser station only when the all-clear signal is given that the lasing area is empty—to avoid crashes. If the parts are not all the same, perhaps a bar code reader is needed to identify the part number and adjust laser settings accordingly.

Lasers run on electricity. A typical laser tool in the United States would run on 208 V, single phase or three phases, with usually about 30 Å per leg. A number of micromachining lasers are single phase and can run on only single-phase power, but if a chiller or larger motors are integrated into the AC string, three-phase power might be needed. Some lasers do require a lot of three-phase power, so in this case either a larger breaker, like 50 Å, or a higher voltage is used, depending on what is available on the shop floor.

Most lasers also need to be cooled and, further, water cooled. Some shop floors and labs have in-house cooling water, but most do not and a chiller is required. For small lasers, a chiller may not add a lot of heat load to the room, but bigger lasers may. There may be a need to use a water–water chiller rather than a water–air chiller if this is the case. Some lasers are very particular and need to be controlled to within a few degrees of their optimum operating temperature. Also, other components of the system may need cooling; for instance, larger aperture galvanometers (30 and 50 mm) usually require cooling because of the high-power motors used to move the larger mirrors.

If flexible materials are being processed, a vacuum chuck and pump may be required to keep the part flat. Some labs have house vacuum, but if not this pump is provided with the laser workstation. Pump size depends on the application. Additionally, there is some amount of effluent from the laser processing and this should be carried away, if possible, so that it does not pose a health problem to the workers or cause problems with the laser tool (dirty optics, etc.). Some labs have house ducting, but the tool may need to be fitted with an air scrubber if the lab is not so equipped or if more vacuum or filter power is needed. Toxic materials like Ni dust, BeCu, and some volatile organics must be handled with even more stringent precautions.

If a laser is used that requires replacing gas in the oscillator or if assist gas is used, a gas-handling system will be needed. Hazardous gases like F_2 and HCl are used in excimer mixes and, not only must the gas delivery be appropriate, but also there must be a mechanism in place to dispose of the waste.

Vented gas cabinets and chained racks are common, with alarmed sniffers used for harmful gases. Even benign process gases like N_2 and He must be handled carefully as there are dangers associated with handling large, heavy, high-pressure containers. The first is oxygen depletion in a closed room if the bottle is opened. The second is handling the high-pressure cylinders, which can be deadly if dropped and the stem is broken. Finally, every facility should be equipped with CDA. Air compressors should be capable of delivering at least 80 psi continuously and this air should be processed through a scrubber to remove any moisture and other contaminants. CDA can be used as a process gas, but it is also used to drive automation pneumatics: part handling, opening and closing doors, etc.

5.3 Part Viewing Systems

Choice of an imaging system depends on the dimensions of required optical parameters and practicality. Two types of imaging systems are common: long working distance and microscope.

5.3.1 Long Working Distance Optical Systems

Long working distance (LWD) setups generally use objective focal lengths greater than 50 mm with objective distances up to 2 m or more (Table 5.1). Working distances are correspondingly large. The depth of field (DOF) is defined as the distance between the nearest and farthest objects in a scene that appear acceptably sharp in an image. Theoretically, a lens can precisely focus at only one distance at a time, with a gradual decrease in sharpness on each side of the focus. In photography, it may be desirable to have the entire image sharp, so a large DOF is appropriate. In other cases, a small DOF may be more effective, emphasizing the subject while deemphasizing the foreground and background.

For laser viewing, DOF is typically defined as the range of distance within which the focus does not change by more than a certain percentage, usually less than 10%. DOF is determined by the camera-to-subject distance, the lens focal length, and the lens clear aperture. Demagnifications are somewhat low, but the field of view (FOV) is correspondingly large, allowing a larger area on target to be viewed. However, the resolution of the image is not great and it can be problematic viewing very small images.

In LWD laser imaging systems, the laser objective lens usually cannot be incorporated as an optical element of the part viewing optics because LWD objectives are typically not chromatically corrected for the visible spectrum. The general approach in this case is to employ a completely separate cam-

TABLE 5.1

Characteristics of LWD Lenses

	Plano-convex (singlet)	Corrected (doublet)	Four Element (corrected)	Multielement (telecentric)
Resolution	>10 μm	~5 μm	~2 μm	~2 μm
Field size (mm)	~10 mm	~10 mm	~5 mm	Up to 25 mm
Complexity	Very low	Low	Moderate	High
Cost	Very low	Moderate	Moderate	High
Losses	2% to 5%	5% to 10%	5% to 10%	>20%
Notes	Very inexpensive Barrel distortion	Dual wavelength operation Moderate distortion	Barrel distortion	Dual wavelength operation Good depth of field

era viewing system. Table 5.2 summarizes advantages and disadvantages of LWD systems.

One approach is to position the viewing system slightly off axis to avoid obstruction of the laser beam (Figure 5.1). This permits high-contrast viewing with a camera and zoom lens assembly. The main disadvantage to this setup is the parallax inherent to off-axis viewing, although corrected optics that minimize angular distortions are commercially available for off-axis viewing.

Another approach is to use a mirror with a central opening to permit on-axis viewing (Figure 5.2). The laser beam is aligned to pass through this opening without hitting the mirror and parallax errors are eliminated. A disadvantage to this setup is a loss of contrast in the part image due to the central obscuration in the viewing mirror. Furthermore, the laser beam must be confined to the opening in the mirror.

A third option is the on-axis, off-line viewing setup illustrated in Figure 5.3. In this case, the work piece is slewed a known distance from the online beam axis to the viewing axis. This distance can be known precisely using high-precision X–Y tables and the offset can then be integrated into alignment calculations. It is possible to use two or more cameras with different X–Y offsets—for instance, one low-magnification

TABLE 5.2

Advantages and Disadvantages of LWD Systems

Advantages	Disadvantages
Small numerical apertures with large field sizes	Small demagnification factors
Large depth of field	Long optical path lengths
Resolution to 2 μm	On-target viewing problematic
Low optic losses	Large support structure required

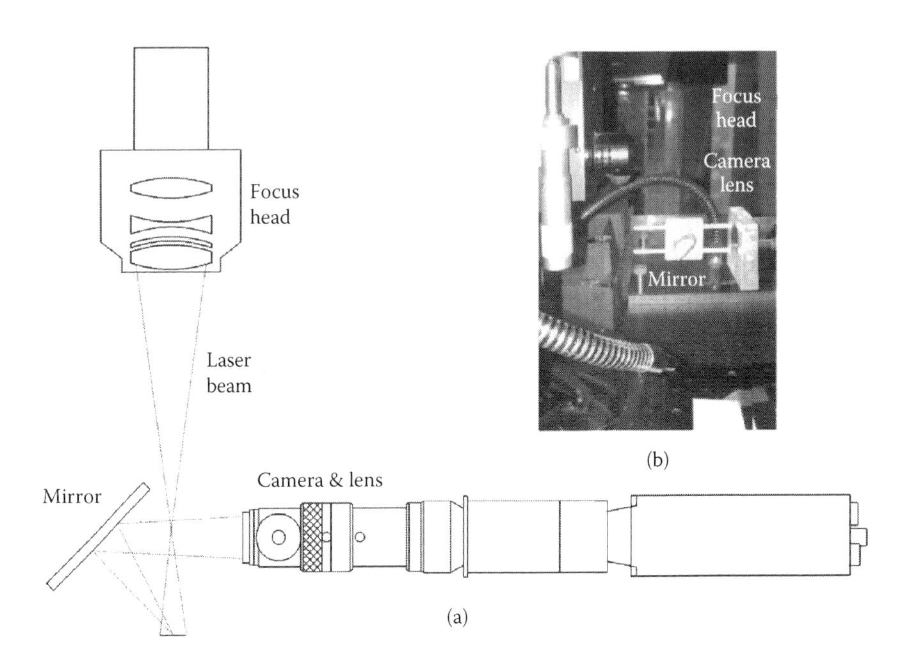

FIGURE 5.1
Off-axis camera arrangement.

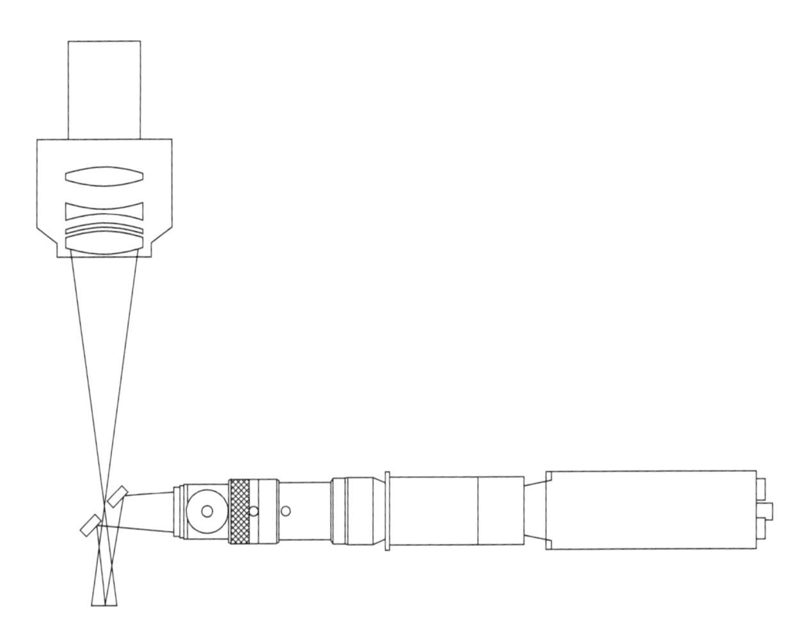

FIGURE 5.2
On-axis camera arrangement.

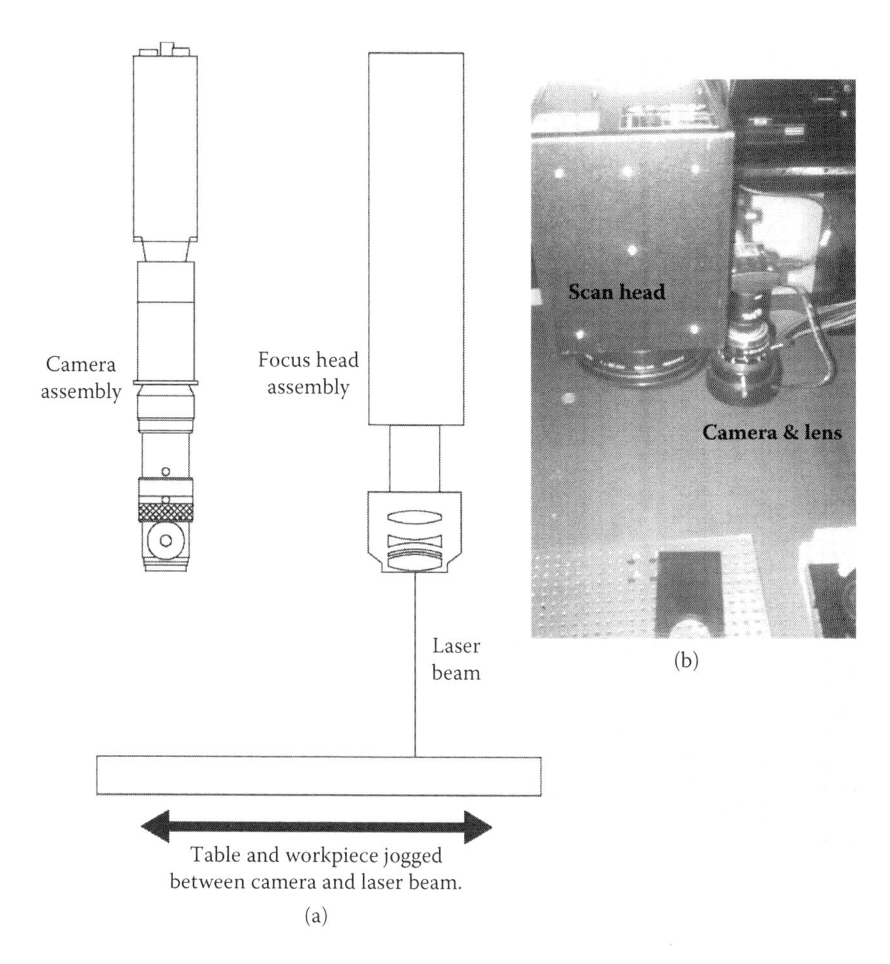

FIGURE 5.3
On-axis, off-line camera arrangement.

camera for global alignment and one high-resolution camera for more precise alignment.

Still another part viewing approach is the on-axis, online viewing setup shown in Figure 5.4. This setup is the most costly. The dielectric optic in the center is UV coated for reflection on the bottom side and visual coated for transmission on top. This viewing setup is particularly useful for microscope viewing; however, the field of view is very small because of the high magnification of the objective. Online viewing gives the fastest feedback since no table jog is used, but viewing through the laser lens also limits the field of view. In addition, for the most part, the process cannot be watched in real time since white light intensity from the plasma and plume overloads the camera.

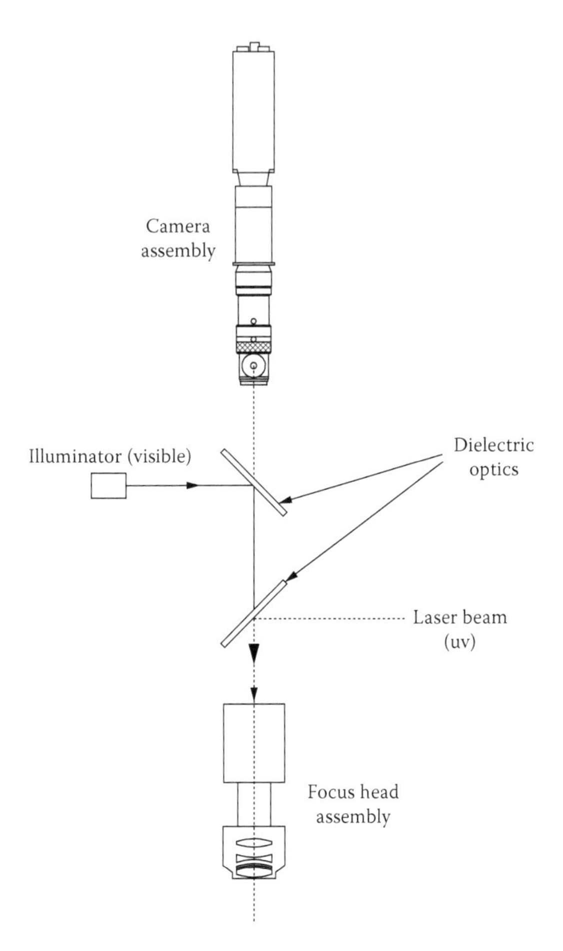

FIGURE 5.4
On-axis, in-line camera arrangement.

5.3.2 Microscope Imaging Systems

Microscope imaging systems have short objective focal lengths up to 30 mm. Working distances are on the order of a few millimeters and the corresponding depths of field are a few microns. Very high demagnifications with factors of tens to hundreds can be achieved, but the beam utilization can be poor. Microscopes also require more complex illumination schemes. It is typical to use through-the-lens part viewing with a microscope objective as achromatic lenses permit simultaneous imaging of, for instance, UV and visible light, thus allowing on-target viewing during processing, as seen in Figure 5.5. A disadvantage to microscope imaging is the short focal length. Very little space exists between the lens and the workpiece. Table 5.3 lists some advantages and disadvantages of microscope imaging systems.

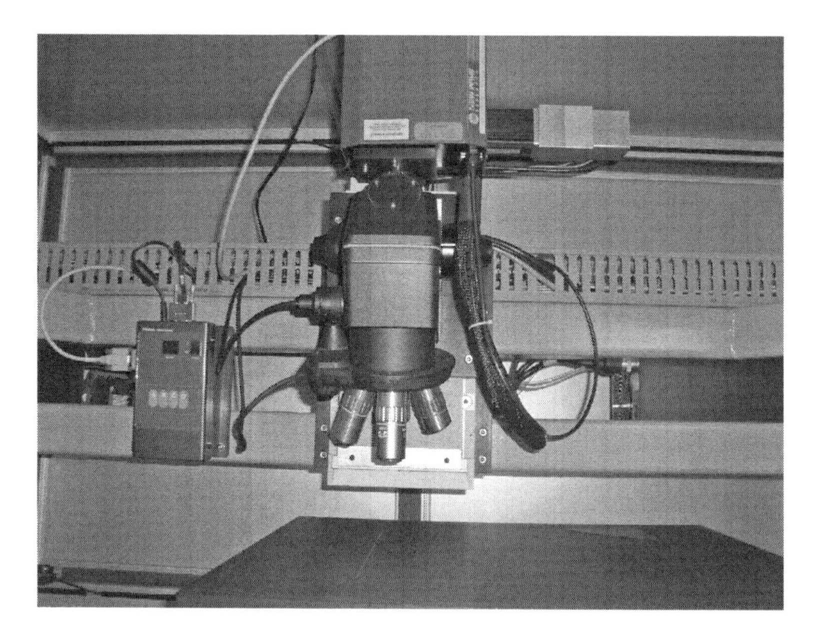

FIGURE 5.5
Microscope objectives and part viewing.

TABLE 5.3
Advantages and Disadvantages of Microscope Viewing

Advantages	Disadvantages
Large numerical apertures	Limited field of view
Short optical path lengths	Short depth of field
High-magnification on-target viewing	High cost
Resolution to <1 μm	High optical losses
Large demagnification factors	Complex illumination required

5.4 Motion Control

The main components in a motion control system are the motor, transmission, bearings, feedback, power/control electronics, and software. All are used in laser machining processes.

5.4.1 Motors

Stepper motor systems are low cost, rugged, compact, and simple in design and require little or no maintenance. They are highly reliable and can be high resolution (<10 μm) when microstepping is incorporated. They have a stiff stationary holding torque and a high operating torque and are ideal for low-speed applications. Some disadvantages of stepper motors include their limitations in positioning accuracy, high operating noise (vibrations), electric current consumption, speed, and potential to stall. The current required for a stepper motor to hold position produces thermal gradients that can impact work point accuracy. In general, stepper motors are less than ideal for contoured motion and constant velocity due to vibration caused by the stepper motor's high pole count and "cogging." Stepper motors are typically run "open loop," or without any feedback. Any system that runs open loop will need some sort of error checking to make sure that the motor does not stall.

Servo motor systems are highly accurate when used with high-accuracy, high-resolution position feedback. Both linear and rotary servo motors are available. In general, a servo motor will run cooler than a stepper motor. As long as the control system does not detect any error, there is no electrical current consumption. When tuned correctly, the motion is smooth. They are ideal for both high- and low-speed applications. Disadvantages to servo motion systems are their complexity and higher cost.

Linear motors produce a linear force along their length instead of a rotary torque, as with steppers and servos. A Lorentz-type actuator is the most common, where the applied force is linearly proportional to the current and the magnetic field. Linear motor systems are highly accurate when used with the proper feedback. They have smooth motion and are extremely fast. They can also be scaled to large-stage sizes.

5.4.2 Transmission Methods

There are two ways to apply force or torque to the payload: *mechanical transmission/gearing* or *direct drive*. Rotary-to-linear transmission methods typically involve a ball screw, belt drive, or rack and pinion. Mechanical transmissions for rotary stages typically use a worm drive, cam, tangent arm, belt, or planetary/harmonic gear. Using these types of mechanical transmission can provide an increased mechanical advantage, although any type of mechanical transmission will add friction, vibration, and backlash.

A direct-drive system does not involve any gearing. The lack of gearing minimizes friction, backlash, and vibration. In a direct-drive system, the bearings are the only contacting element. In some cases, even the bearings are noncontact. Direct-drive systems are ideal for contoured motion, high- and low-speed scanning, and very small incremental motions (steps). The disadvantage of direct-drive systems is their lack of mechanical advantage. Vertical linear motor axes and unbalanced rotary axes may require the use of a counterbalance or counterweights.

5.4.3 Bearing Technology

There are many different bearing options from which to choose. Used in research and industrial laser processing machines, *linear bearings* include square rail/linear motion guides, crossed roller bearings, and air bearings. Linear motion guides utilize recirculating elements that come in and out of contract with the guideway. These recirculating elements are usually ball bearings. Mechanical noise/vibration occurs as the balls enter and exit the loading zone. Friction between the balls and guideways, sphericity of the balls, and surface quality/straightness of the guides also limit performance. Linear motion guides are capable of long stroke lengths and high speed. A ball bearing can be thought of as a point of contact.

Crossed roller bearings are composed of rolling cylinders between a stationary and movable bearing way. The rollers are placed in a row with their rotation axes alternating 90°. Unlike linear motion guides, the contacting elements in a crossed roller bearing are always in contact with the bearing ways. Crossed roller bearings are limited in stroke length due to the moving guideway. Friction between the rollers and bearing ways, roller tolerances, and surface quality/straightness of the bearing ways can limit performance. The line contact of the rollers provides excellent load capacity and stiffness.

Air bearings are used in the highest performance linear and rotary motion stages and have substantial advantages over conventional contact bearings. The stiffness of the air bearing is proportional to the bearing surface area. Air bearings have an inherent averaging effect due to the pressurized air film that fills voids and surface imperfections. The result is excellent geometric performance (minimizing errors arising from straightness, flatness, yaw, pitch, and roll). In an air bearing linear stage, those unwanted errors can be at least two to five times lower than the best contact bearing stage.

Air bearings also provide the least vibration, making them the ideal choice for high-accuracy scanning and contouring. The advantages become more pronounced as the resolution increases. There is a tremendous advantage in having the three critical components of the stage—the guideways, the motor/actuator, and the position feedback—all completely noncontact. In contrast to conventional stages, air bearing direct-drive stages have no contact and no friction, so they develop a force in an extremely linear and predictable manner. Friction is a highly nonlinear effect and degrades the performance of servo control loops. The absence of friction from air bearing direct-drive stages permits much higher static and dynamic performance to be achieved and allows a substantially shorter settling time.

Air bearings do have some disadvantages. They do not have any damping. Also, there is a modest increment in system cost and complexity for air bearings over conventional stages due to the need to provide a supply of clean, dry air.

5.4.4 Other Motion Elements

Other motion elements include goniometers, tilt stages, z-axis lift stages, and rotary stages. A goniometer is an instrument that either measures an angle or allows an object to be rotated to a precise angular position. A positioning goniometer or goniometric stage is a device used to rotate an object precisely about a fixed axis in space. It is similar to a linear stage; however, rather than moving linearly with respect to its base, the stage platform rotates partially about a fixed axis above the mounting surface of the platform. Positioning goniometers typically use a worm drive with a partial worm wheel fixed to the underside of the stage platform meshing with a worm in the base. The worm may be rotated manually or by a motor as in automated positioning systems.

Highly accurate motion systems require a base and environment isolated both thermally and vibrationally. A motion base made from a heavy block of granite is the accepted way to do this, although new materials are available that can be easily machined and still provide the necessary rigidity. Normally, the stacked motion elements and the beam delivery are on the granite. This is mounted to the base only (extruded Al or foam-filled weldment) with a three-point contact system so that the rest of the frame and enclosure are completely isolated from the working area. Figure 5.6 shows a gantry-style motion platform where the upper axis sits on two stages working in tandem (one on each side of the lasing area). This configuration costs more because of the additional motion axis, but it can save precious floor space, especially for large travel stages and in high-cost environments like clean rooms.

5.4.5 Power/Control Electronics

Brushless servo, DC brush, and stepper motors all require an amplifier or drive to power them. Amplifiers can vary in their complexity, but for simplicity, only those that allow position or velocity control are discussed here. Torque is proportional to current and velocity is proportional to the amplifier's bus voltage. Actual torque and speed characteristics for each motor are determined by the motor's winding characteristics.

Two types of amplifiers are used in motion control: linear and PWM (pulse width modulation). *PWM amplifiers* are the most widely used due to their efficiency, high power, and low cost. They regulate current through the motor's windings by taking the bus voltage and switching it on and off at a certain duty cycle. The more current required to meet the controller's demand, the higher the percentage of controller "on" time will be. Typical PWM switching frequencies are 10 kHz or higher. PWM amplifiers are not appropriate for all applications. Due to their high switching frequency, they generate unwanted conducted and radiated noise at the fundamental switching frequency and harmonics. PWM amplifiers also have a "dead band" and do not behave linearly at very low current levels. This can be problematic for high-precision applications. The dead band and nonlinearity can result in increased dynamic errors during direction reversals and contouring. PWM amplifiers also exhibit high-frequency ripple currents that can influence in-position stability and velocity jitter.

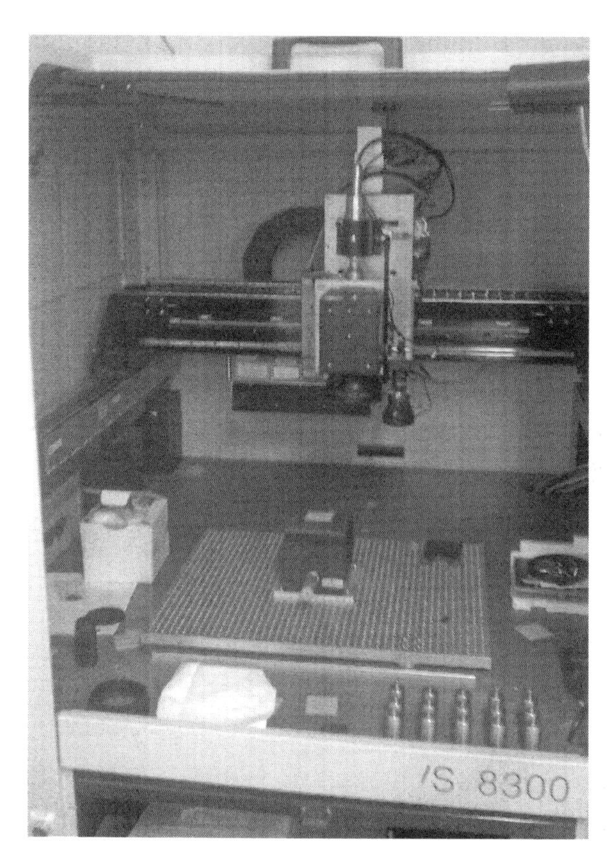

FIGURE 5.6
Gantry-style motion platform.

When *linear amplifiers* are compared to their PWM counterparts, the former are more expensive, generate more heat (they are inefficient), and take up more space. Linear amplifiers are chosen when the application requires the best possible constant velocity scanning, <50–100 nm in-position stability or extremely tight contouring/dynamic accuracy. Linear amplifiers apply only the amount of current/voltage to the motor that is required by the controller. They do not switch a bus voltage on and off at high frequencies, so they produce nearly zero conducted and radiated EMI (electromagnetic interference). The best linear amplifiers behave very linearly near zero current and the result is improved in-position stability, contouring, and scanning. Another limitation is that, due to their design, linear amplifiers cannot absorb as much energy generated by the motor when it decelerates large inertial loads or when it is back-driven.

Amplifiers receive their command signal from a controller or trajectory generator. Many amplifiers accept a command signal from a controller that is a clock and direction signal, ±10 V torque, or velocity command. Many of today's amplifiers are digital smart amplifiers that actually close the control loops on the amplifier and receive commands via a distributed control network. Distributed control architectures simplify system wiring and are highly scalable.

5.4.6 Controllers and Motion Software

The motion controller must execute the application program and generate the motion trajectory. For laser processing applications, the motion controller must have many of the same features that a CNC (computer numerical controller) has. Cutter tool compensation (beam width or kerf), multiblock look ahead, acceleration limitation, retrace, coordinate transformations, and G-code functionality are some of these features. It is also ideal to have the controller directly linked to the laser to ensure that a controlled laser dosage is applied to the part. When the motion system enters and exits a corner or radius, it must decelerate and then accelerate out of that radius. If the laser is applying a constant output regardless of velocity, too much dosage will be applied to the part, resulting in heat-affected zones and poor cut quality. For right angles and very small radii, it is not possible for the motion system to provide constant velocity.

The motion controller receives high-level instructions from the computer, which interpolates profiles and controls the drive system. Output to stepper motor drivers is a step signal and a direction signal for each axis. Microstepping can give up to 50,000 steps per motor revolution. Usually there are several velocity profile options and they can be run open loop or closed loop with encoder feedback. Stepper motor drivers control adjustable electrical current in the windings of each motor. Rotary motors are used with a lead screw drive for linear positioning.

The motion controller resides in the computer and uses encoder feedback to position motors. Many velocity profile options with feedback are available. The output signal is usually a voltage (–10 to +10 V) proportional to the required motor speed. Servo amplifiers are usually compatible with two- or four-lead motor hookups and have adjustable current output. Servo motors are two or four leads with current ratings based on torque requirements. Several types of encoders are used, including linear and rotary (physically attached to the motor). They provide 5 V square wave pulses directly to the motion controller card and can achieve accuracies of <1 μm.

In addition to the motion elements, the basic architecture of a motion control system also contains other elements. A motion controller is used to generate set points (the desired output or motion profile) and close a position and/or velocity feedback loop. A drive or amplifier is used to transform the control signal from the motion controller into a higher power electrical current or voltage that is presented to the motion element. "Intelligent" drives can close the position and velocity loops internally, resulting in much more accurate control. One or more feedback sensors, such as optical encoders or Hall effect devices, return the position and/or velocity of the motion element to the motion controller in order to close the position and/or velocity control loops. The mechanical components transform the motion of the actuator into the desired result and can be used to drive gears, screws, belts, linkages, linear conveyors, and rotary bearings. The interface between the motion controller and the drives that it controls is very critical when coordinated motion is required, as it must provide tight synchronization. Common control functions include

- Velocity control
- Position (point-to-point) control (the several methods for computing a motion trajectory are often based on the velocity profiles of a move such as a triangular profile, trapezoidal profile, or an S-curve profile)
- Pressure or force control
- Electronic gearing (or cam profiling; the position of a slave axis is mathematically linked to the position of a master axis)

The most common interface is the use of a computer or PLC (programmable logic controller). Computers allow much greater flexibility and ease of use, but PLCs are fast, robust, and reliable. In many cases, a laser tool incorporates both, especially if a large amount of automation is used, so that the PLC handles routine moves and commands and the computer handles more selective activities.

5.5 Part Handling

The simplest part holder is a flat vacuum chuck that draws through predrilled holes or pores in ceramic or metal mesh and holds thin, flexible material flat. Enough suction is needed to have uniform pull over the entire surface; big chucks sometimes have vacuum ports on multiple sides. Figure 5.7 shows two kinds of vacuum chuck: a hollow aluminum chuck with ¼ in. holes drilled on 1 in. centers and a porous ceramic vacuum chuck. The aluminum chuck is standard, but since the vacuum does not cover the entire surface, it creates three problems. First, some areas do not get vacuum at all, so small parts will not be held flat unless they are directly over a drilled hole. Second, very small parts can get sucked right through the holes. Third, thin materials will dimple at the hole.

Porous ceramic chucks eliminate these problems by using a piece of ceramic filter that is ground flat and parallel with 20 μm pores that cover the entire surface. High-power lasers will eventually eat into the ceramic over time, but they can be resurfaced if needed and, for low-power lasers, they can last through years of use without degradation. Vacuum chucks can have

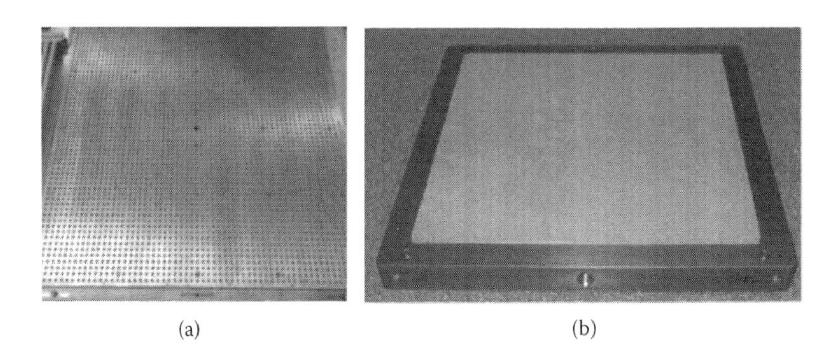

(a) (b)

FIGURE 5.7
(a) Standard Al and (b) porous ceramic vacuum chucks.

several zones that can be turned off and on depending on the part size and can come in all shapes including square, rectangular, and round. Figure 5.8 shows a four-zone chuck and a round chuck.

Other tooling can be made specific to any job. For instance, when laser drilling very small and precise holes in molded plastic parts, it is extremely important to focus on the front surface. This cannot always be accomplished by nesting the body because of the irregularities of the molding process. In this case, spring-loaded positioners can be used that butt the front surface of the drill area against a known stop positioned accurately relative to the focus of the laser. Figure 5.9 shows such a front surface locating mechanism. Another simple holding method for round parts is a "V" block (Figure 5.10), which can be used to maintain focus—for instance, while drilling a catheter along its length. Holes drilled along the length provide a vacuum for assuring that the part is held firmly during processing.

In production environments, roll-to-roll devices, conveyors, and robots are also used to move parts on and off the laser. In fact, part handling frequently becomes as critical in designing a production system as optimizing the laser

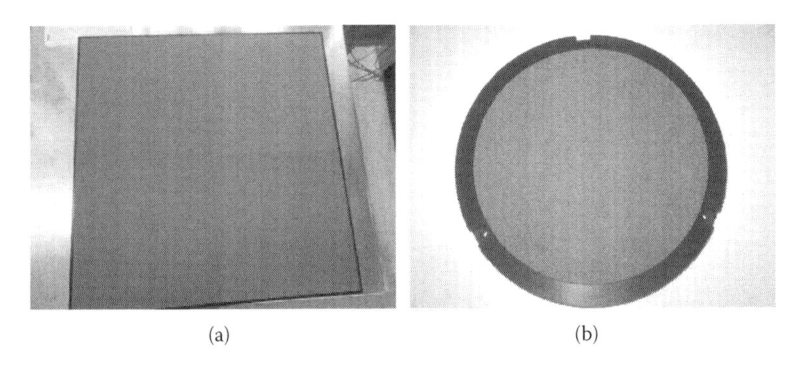

(a) (b)

FIGURE 5.8
(a) Four-zone and (b) round vacuum chucks.

FIGURE 5.9
Front surface location part holding.

parameters. Figure 5.11 shows a roll-to-roll system developed to drill holes in polyimide film using multiple galvo heads. The left side roll is the unwind side, including the tension rods and dancers; after laser processing, the film is taken up on the right side. The process is fully automated, so when a roll is finished, the system goes into standby and an alarm is sounded to alert the operator to change rolls.

Figure 5.12 shows a laser processing station with a conveyor-type load and unload. This system was made for processing glass of different thickness, width, and length, so the conveyor was designed to accommodate a variety of glass sizes automatically. Sensors placed along the travel path track the movement of each glass panel through the tool so that there are no crashes. The system is shown in a stand-alone configuration where an operator loads and unloads each part, but such a tool can be integrated into a fully automated production line with handshaking between stations before and after the laser processing chamber. Note also that the flat glass enters and exits through small slits in order to meet eye safety requirements.

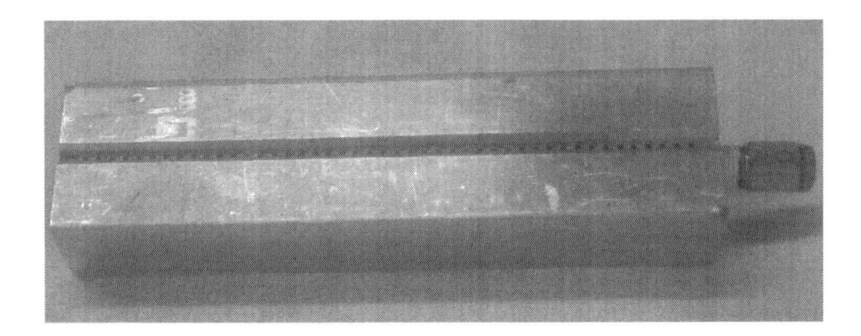

FIGURE 5.10
"V" block part holder.

FIGURE 5.11
Roll-to-roll part processing.

5.6 Laser Support Systems

Laser tools require a few facilities in order to operate. The first is electrical power. How much power is needed depends on the size and power requirements of the laser itself and whether it needs a chiller. In the United States, 208 V, single- or three-phase power (60 Hz) with about 30 Å per leg is used, with 110 V legs operating the small components (motors, computer, amplifiers, etc.). If the laser uses a lot of electrical power, like some larger excimer lasers, 50 Å breakers or higher operating voltages might be required. All

FIGURE 5.12
Conveyor system for glass processing.

electrical power is brought in through a central AC distribution box with a cutoff switch that kills all power to the tool.

All lasers require cooling because the wall plug efficiency, even for fiber lasers, is not 100% and the excess is dissipated as heat. If air or convection cooling is used, only in very low-power lasers, the additional heat load to the room must be accounted for. Water cooling is more typical and is used to cool the laser head, possibly the power supply, and sometimes other components as well. If water cooling is used, a temperature control good to about ±1°C is needed for some lasers, although others are much more forgiving. In addition, an in-line flow sensor is used to assure that there is always a flow and to shut power to the laser off if no flow is sensed. Each laser has its individual preference (distilled water, deionized water, etc.), but the fluid used should be clean and filtered to avoid contamination of the laser head.

Laser systems frequently use gases. Excimer and CO_2–TEA lasers require constant fills from a gas bottle. Especially in the case of excimer gases, gas bottles must be kept in a certified gas cabinet, perhaps incorporating safety valves for toxic gases. Assist gases are also used, especially during IR processing, to cool the parts and help remove effluent. Many types of assist gases can be used (this topic is explored in detail later), but at a minimum, CDA at a pressure of about 80 psi is needed and can be used for assist and also to drive pneumatics.

Finally, precision lasers require a controlled environment if they are to work properly over extended periods of time. Temperature and humidity control are almost mandatory and a clean room environment is preferred (class 100,000) in order to keep the optics clean.

5.7 Software

Software coordinates the operation of the laser, motion control, and workstation peripherals. It is at the heart of any good laser materials processing system and can make a world of difference in system usability, maintenance, and safety. The main purpose of software is to control the laser tool to make a complex, high value part efficiently. Every laser system manufacturer has its own ideas about what software should look like, how the graphical user interface (GUI) should look and feel, and what level of control and access should be allowed for different users. Software development is an ongoing process because computers and operating systems are constantly changing and evolving; this influences not only the program, but also how other hardware devices may interact in different operating environments. It is further complicated by the fact that there are many different lasers, different laser manufacturers, different motion control platforms, galvanometer platforms, camera drivers, etc.

Consider a program written in C++ with .NET Framework 3.5. (Visual Basic is becoming less practical as it is not supported to the same level as other programming languages.) Figure 5.13 shows a top level diagram. The *initial form* (window) is displayed while the software initializes the hardware and sets up the main window. The *main form* is displayed for most of the software's run time and allows user interaction—setting parameters, loading files, viewing the live video feed, and so on.

From the *main form*, a number of other forms can be called. In this case, there is the *laser selection* form, which allows selection of the system's laser type; the *job editor* form, which allows the user to create control programs that coordinate stage movement with laser processing; and the *about* form, which will display version information to the user. When the user is ready to shut down the system, the *main form* is also the one to call the *shutdown* form, which is the window that is displayed as the software closes out the *main form*, cleans up after itself, and shuts down the hardware.

In this example, the *main form* contains many different modules, each with its own task to complete. Figure 5.14 shows a module-level diagram. The *stage*

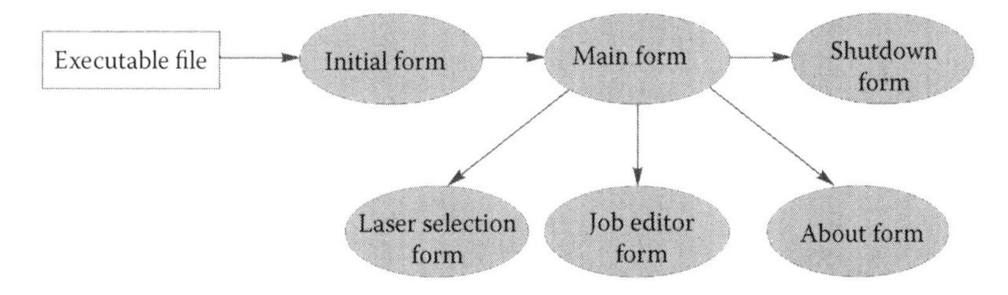

FIGURE 5.13
Top-level diagram of PMI C++ program.

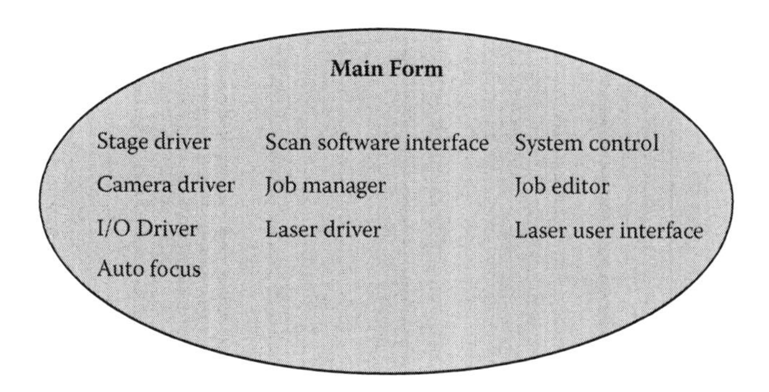

FIGURE 5.14
Module-level diagram of PMI C++ program.

driver module takes care of implementing motion control commands such as moving a stage to a specific location. The *laser driver* will handle communications with the laser, each of which has its own unique features. *Laser user interface* is a module that contains the controls available to the user on the *main form*, based on which laser is currently configured. *Camera driver* handles the image feed coming from the currently installed camera and its specific parameters.

Making software modular is very beneficial in building complex systems such as these. If needed, one component can then be easily swapped out for another. Perhaps a piece of hardware has changed, or an upgrade has been made to an existing module. In addition, having discrete components, each in its own isolated block of code, allows for multiple sections of the code to be worked on simultaneously without fear of cross-contaminating various versions of the main code base.

The introduction of multiple threads has also allowed processes to occur in parallel to each other. Care needs to be taken to implement correctly, but it can improve responsiveness and accuracy of different software functions, such as quickly checking for interlock conditions. It can also offload intermittent functions, such as updating the displayed stage position, so that they do not interfere with more important processes.

All of these things working together make up the main body of the software, and its modular design allows for future expansion and addition of new features. This functionality can be seen in several pieces of software in the form of "plug-ins," which can even allow the end user to design new functions (should he or she choose to allow it).

Figure 5.15 shows a typical *main form* screen, which shows the preceding principles in action. The main screen is the interface of the software that allows the user to interact with the system while hiding all the functions that are happening beneath the surface.

The video feed seen in the upper left can be set to "live" video or individual "snapshots" in the "vision control" frame. These images and the settings

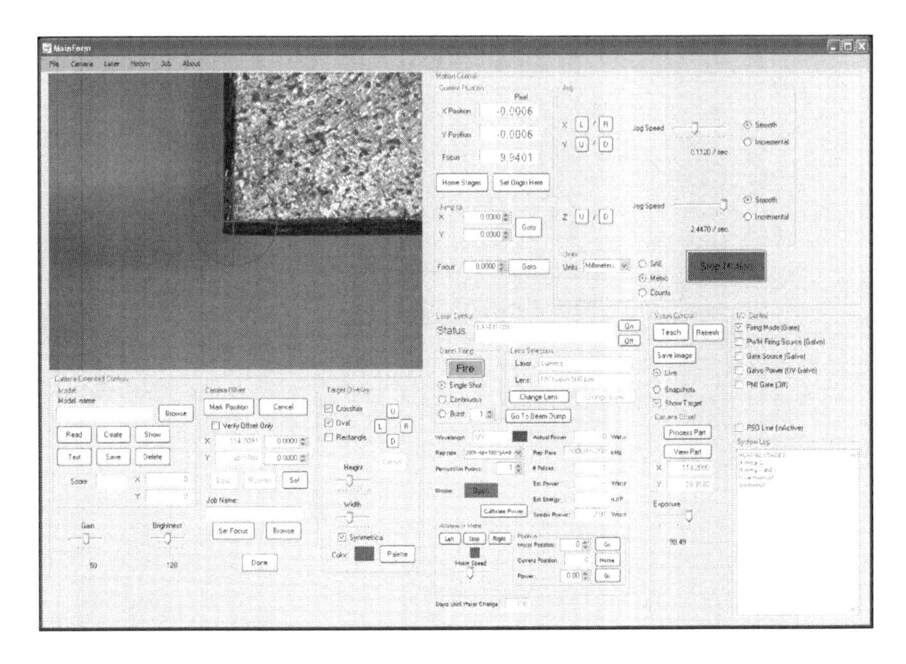

FIGURE 5.15
PMI main screen.

that determine their appearance are controlled by the aforementioned *camera driver* module. The "overlay target" can also be toggled on and off and the exposure setting adjusted. The stage can be moved from the camera to the laser and back by making calls to the *stage driver* module.

More camera options are available as well by clicking the "camera" menu option, which displays the "camera extended controls" underneath the video box. This set of controls allows the user to set gain and brightness settings of the camera, to adjust the properties of the target overlay (in this case, a purple crosshair), to manage and test models that are used with pattern matching image recognition, and to manage the X–Y offset between the camera and the laser. Other controls, such as the job manager, can be set to occupy the same space by clicking the associated menu item at the top.

The "motion control" frame has many *stage driver* interactions available, such as moving the stage with buttons in the "jog" frame. The current relative position of the stage is shown in the "current position" frame, which also has controls to allow the user to home the stages and set new origin positions. The "jump to" frame will allow the user to enter a specific relative X–Y or Z position and move the stage to that location. "Units" are selectable. The big red "stop motion" button is there for safety.

The "I/O control" frame handles the hardware options that are able to be toggled on and off, such as switchable power outlets, selected galvanometers, or laser trigger sources. Some systems have dedicated I/O hardware, while others simply use the onboard I/O available on the motion controller.

The *laser UI* module also has its interface in the main window. This is a good example of a module that may be unique to every system, since every laser has its own set of features and controls. For example, although some lasers allow a wavelength change, most do not. There are common controls among lasers, such as the ability to set the lens type and calibration file in a system that is using a galvonometer, as seen in the "lens selection" frame. Most lasers will allow external triggering or gating, so the "direct firing" frame would likely be seen in most systems.

There may even be interactions between some of these modules, such as the case where a stepper motor signal from *stage driver* is being used by *laser driver* as a trigger signal. The user will never see these interactions, however, since they are working together invisibly beneath the surface.

5.8 Safety

Safety is of paramount importance! All systems involved in a laser materials processing tool should address any and all safety concerns before the key is turned on. The many aspects to keeping the lab safe are addressed in detail next.

5.8.1 Laser Safety

The first thing most people think about when they think of laser safety is the eye. High-intensity ultraviolet and infrared light can be hazardous to the eyes and skin. Visible to near-infrared light from 400 to 1400 nm can injure the retina. Beyond 1400 nm, the injury will be to the cornea. Corneal injuries occur from the mid-ultraviolet wavelengths from 180 to 315 nm. Wavelengths from 315 to 390 nm may produce opacity in the lens of the eye.

All manufacturers are required by law to design and certify laser systems in compliance with the US Code of Federal Regulations, parts 1040.10 and 1040.11. The CDRH (Center for Devices and Radiological Health) branch of the FDA has oversight. Class I, II, III, and IV laser safety ratings are based on the level of radiation that is accessible by humans during normal operation of the equipment. Class I laser systems are fully interlocked, which means that the beams are fully enclosed under normal operating conditions, all doors are closed and all windows are chosen so that the wavelength of the laser used is also blocked. If the integrity of the interlock string is compromised, a shutter is closed or a power supply is disabled. This assures that no laser light can exit the enclosure. A key switch and password are used to access a service mode of the laser, useful for aligning and servicing the laser tool. The tool should have safety labels per CFR 1040.10 at appropriate places, and a log should be kept for every tool. *Good record keeping is important for the operation of the laser tool and also for safety.*

Ultraviolet micromachining lasers produce microjoules to millijoules of energy per pulse and operate from a few hundred hertz to a few hundred kilohertz. Infrared lasers have much higher energy per pulse. For ultraviolet lasers and long wavelength infrared lasers (CO_2), low-level exposure to the eye will interact with the cornea and will not continue through to the retina; however, the retina will be affected in the near infrared. Nd:YAG lasers are among the top five lasers that are involved in over 70% of all eye injuries. Using a green 532 nm laser poses potentially even more hazards because this wavelength will pass through the cornea and be focused by the lens onto the retina, resulting in a retinal burn even at low exposure levels. Extreme care must be taken when working with visible light. The skin is also subject to burning, even from diffuse stray light from an ultraviolet laser.

When aligning and using a laser in a Class IV configuration (fully exposed), in addition to laser safety eyewear, long-sleeved shirts should be worn with no jewelry (stray reflections can occur). If skin is exposed to even stray light, sunblock should be used. Every company should have a designated laser safety officer (LSO), and one is required for all Class 3B and Class 4 lasers or laser systems. While there is no federal or state requirement that LSOs be certified, it can help to reduce insurance costs. More details on laser safety, choosing the correct eyewear, and safe operation of lasers can be obtained from the Laser Institute of America (www.lia.org), whose mission is to foster lasers, laser applications, and laser safety worldwide.

5.8.2 Mechanical Safety

Large motion control systems, conveyors, and robots may pose mechanical hazards. Pinch points like closing doors must be labeled and guarded. Mechanical guards or light curtains are used to minimize potential hazards. Closed-toe shoes at least, if not safety shoes, should always be worn in the lab.

5.8.3 Electrical Safety

Lethal high voltages appear in the laser head and power supply. Only qualified personnel should attempt to work with the electrical portion of the laser tool as this area can be potentially the most lethal. Safety covers should be in place on all electrical components. Every tool should have an emergency off (EMO) circuit, which is a large *red* button in a yellow housing that kills power at the AC junction box to every element in the laser tool. Sometimes the computer is put on a battery-powered uninterruptible power supply (UPS) so that a hard crash can be avoided. Figure 5.16 shows the front panel of a laser tool with the interlock key switch and EMO button. Figure 5.17

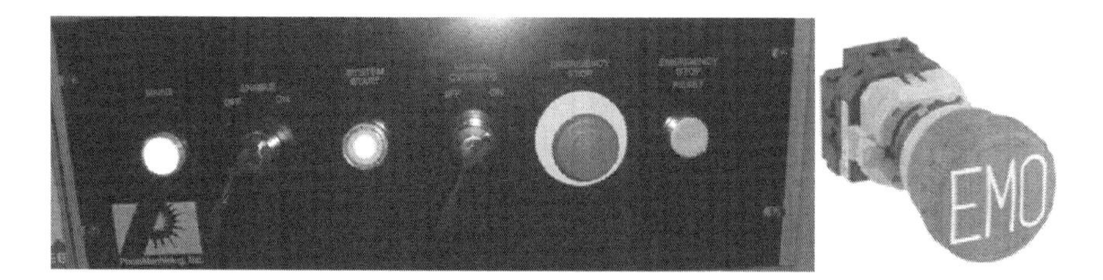

FIGURE 5.16
Front panel with interlock key switch and EMO button.

FIGURE 5.17
Four-color safety light.

shows a four-color safety light. These lights are often used to indicate status; for instance, a red light would alert operators to a potential hazard. These lights are often equipped with sirens that give an audible alert as well as a visual alert.

5.8.4 Materials Safety

Toxic and corrosive fluorine and HCl gases are used in an excimer mixture and provisions have to be made for handling these gases. The concentration in the bottle is only 5%, and in the laser head itself the concentration is about 0.1%. These gases must be changed regularly, and the used gas also contains not only the initial toxic gas, but also halogen-containing compounds formed during laser operation. All gas cylinders need proper handling and storage as they are large and, if the neck is broken, they can become deadly missiles. Even with otherwise benign gases, leaks in small, closed rooms can lead to oxygen depletion.

The materials themselves or the effluent from laser processing may also be hazardous, or at least noxious. Fortunately, only a very small amount of material is removed in laser micromachining. Nevertheless, care must be taken to avoid undue exposure. Some metals, like BeCu and nickel, are hazardous and breathing in or even touching the dust and fumes that are generated should be avoided. Some polymers likewise create effluent that needs to be addressed. A well-ventilated and -ducted laboratory is crucial. Local regulations should be reviewed as well as federal statutes to assure total compliance. Note that in no case can any state have safety standards below those of the federal statutes, and in many cases the state and/or local regulations are much stricter.

6

Discussion of Some Processing Techniques

6.1 Aligning to Fiducials

When running a job on a laser (or other tools for that matter), it is frequently required that the processing tool path be aligned to existing features. As a general statement, the sharper and smaller the alignment feature, the more accurate the alignment will be. Typically, a camera or optically based vision system is used. Assuming high-resolution motion stages and cameras and crisp fiducials, alignment accuracies on the order of microns are possible.

Consider a simple case using a two-dimensional printed circuit board. The number of alignment points defined in the tool path determines the accuracy level to which a panel can be aligned. The global alignment features should be placed as far apart as possible for the best results (Figure 6.1) and in all cases should be outside the laser processing area. If individual components on a panel require very high accuracy, sometimes it is best to use fiducials more local to the components to increase accuracy; however, this also increases cycle time as there are more alignments per panel. One of four levels of alignment is used:

One-point alignment corrects an X–Y offset between an alignment point on the panel and the reference point in the file.

Two-point alignment corrects panel rotation and X–Y offset. The X–Y offset is calculated using an average of the offsets of the two aligned points. The angular correction is derived from the offset angle of the two aligned points. See Figure 6.2.

Three-point alignment corrects for X scaling and Y scaling (stretch and shrink) in each axis. The average offset of the three points is used to derive the X–Y offset. The angular offset is calculated separately in each axis and is effectively an orthogonality (skew) correction.

Four-point alignment corrects for X and Y linear scaling (keyhole scaling) in addition to three-point alignment. It accounts for X as a linear function of Y and for Y as a linear function of X.

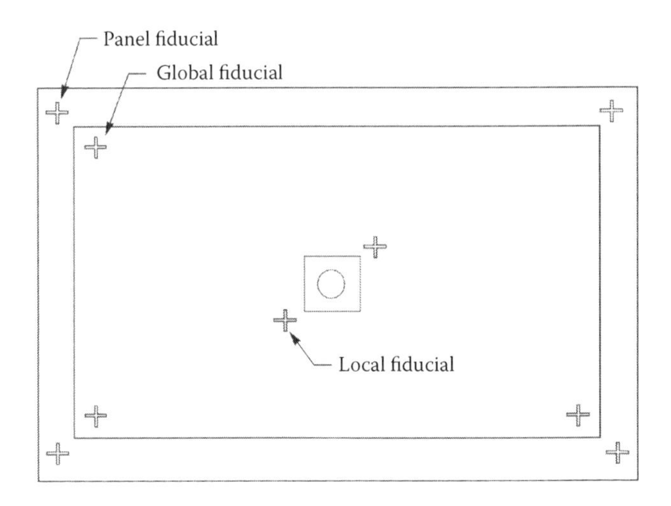

FIGURE 6.1
Local and global alignment.

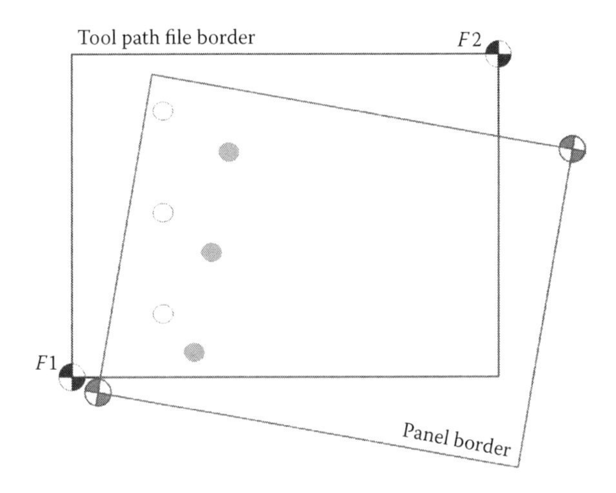

FIGURE 6.2
One- and two-point alignment schematics.

In summary, one-point alignment corrects for linear offset, two-point alignment for linear offset and rotation, and three-point alignment for linear offset, rotation, and stretch/shrink; four-point alignment adds keyhole scaling to three-point alignment. For products that tend to distort during the manufacturing process, such as printed circuit boards, three- or four-point alignment is commonly needed. For stable, rigid products like metals, silicon, glass, and ceramics, one- or two-point alignment is probably sufficient.

For a manual alignment procedure, an operator will watch a monitor and use the X–Y jog to align a set of camera crosshairs to the fiducial. The camera

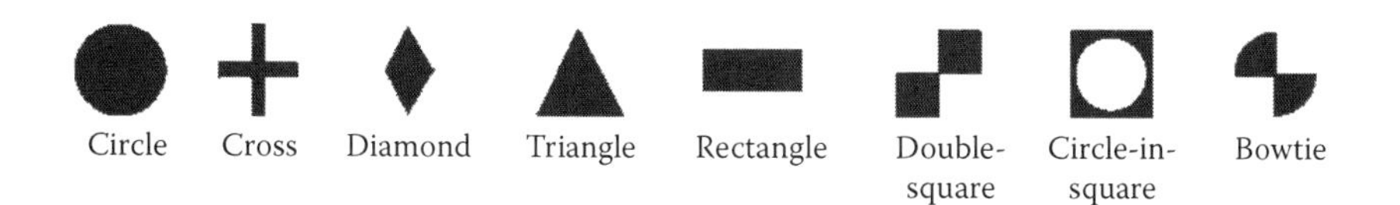

FIGURE 6.3
Different marker types.

is in line with the laser or, more commonly, it is set at a known offset to the laser beam, which is factored into all positional calculations. Once the first alignment is complete, the process is repeated over the number of alignment points used. After the last alignment is made, calculations are made and the tool path is defined with respect to the new alignment and camera offset information. This process can be automated using machine vision, which can be much faster and more accurate than an operator but does add expense to the tool. Automated alignment is particularly valuable when there are high numbers of one part to be made and less valuable in very small lot sizes.

A *fiduciary marker* or *fiducial* is an object used in the field of view of an imaging system that appears in the image produced for use as a point of reference or a measure. One fiducial is usually designated the primary fiducial and is used to adjust the X–Y offset. Secondary fiducials are used for other operations. Frequently, the fiducials are not all the same, so operators or automated vision systems must take into account different types. Figure 6.3 shows several common markers. Some general characteristics of fiducials include the following:

Shape. The optimal fiducial mark depends on the requirement. For instance, the crosshair shape that would be easy for a human to use may not be appropriate for machine vision that uses edge detection, since the line width for both axes needs to be taken into account. A circle would be less prone to error because the positional data include only a single center point. If the machine vision technology is based on creating and then comparing detected fiducials to a model, the shape becomes less important as long as it is repeatable. If the fiducial itself varies from unit to unit, then the resulting detection accuracy will suffer. Some fiducials are a combination, which allows the same unit to work well with different methods of detection. For example, the "bowtie" shape has two rounded edges on the outside to generate a circle, as well as a crosshair in the middle for manual alignment.

Size. Smaller and crisper fiducials give better alignment capability. Minimum diameter is determined by the optical system and the resolution of the motion hardware; maximum size is determined by the field of view of the imaging system. Fiducial marks located on the same part should not vary in size by more than a few percent.

Some assembly equipment is less flexible in its ability to recognize different size marks on the same part.

Clearance. A clear area devoid of any other features or markings should be maintained around the fiducial mark. The size of the clear area should minimally be equal to the radius of the fiducial mark. A preferred clearance around the mark is equal to the mark diameter.

Material. The material will be dictated by the part, but etched metal, organics, raw ceramics, or other materials are encountered. The main points are resolution of mark, permanence, and contract.

Flatness. The flatness of the surface of the fiducial mark should be 15 μm (0.0006 in.) or better.

Edge clearance. The fiducial marks should be located no closer to the edge than 7.62 mm (0.300 in.) (Surface Mount Equipment Manufacturers Association standard transport clearance).

Contrast. Most machine vision recognition systems perform best when a consistently high contrast is present between the fiducial mark and the base material.

When bare material is processed, alignment problems are quite minimal, but when aligning to existing features is undertaken, some care must be taken to place location markers in the correct places for alignment and also to make sure that the marks are of the correct size and contrast. Up-front design can avoid many pitfalls and make the process of alignment go more smoothly than without forethought. Designing parts up front with alignment in mind takes no more time or money, but can greatly influence run times and accuracies on the production floor.

6.2 Laser Drilling Large Numbers of Really Small, High-Aspect Ratio Holes

The term "small" is relative and subject to individual interpretation depending on the application. Considering other complementary technologies and also most laser-based technologies, a hole less than 100 μm in diameter can be considered small. However, in this case the discussion is limited to holes less than 10 μm in diameter and even down to fractions of a micron. To make holes this small, the only real laser option is an ultraviolet (UV) laser: either 355 nm or 266 nm DPSS (diode pumped solid state) lasers or excimer lasers. Also, the UV photons have to be used with the correct optical system and this usually means that a fairly short focal length lens is used to get very small spot sizes on target.

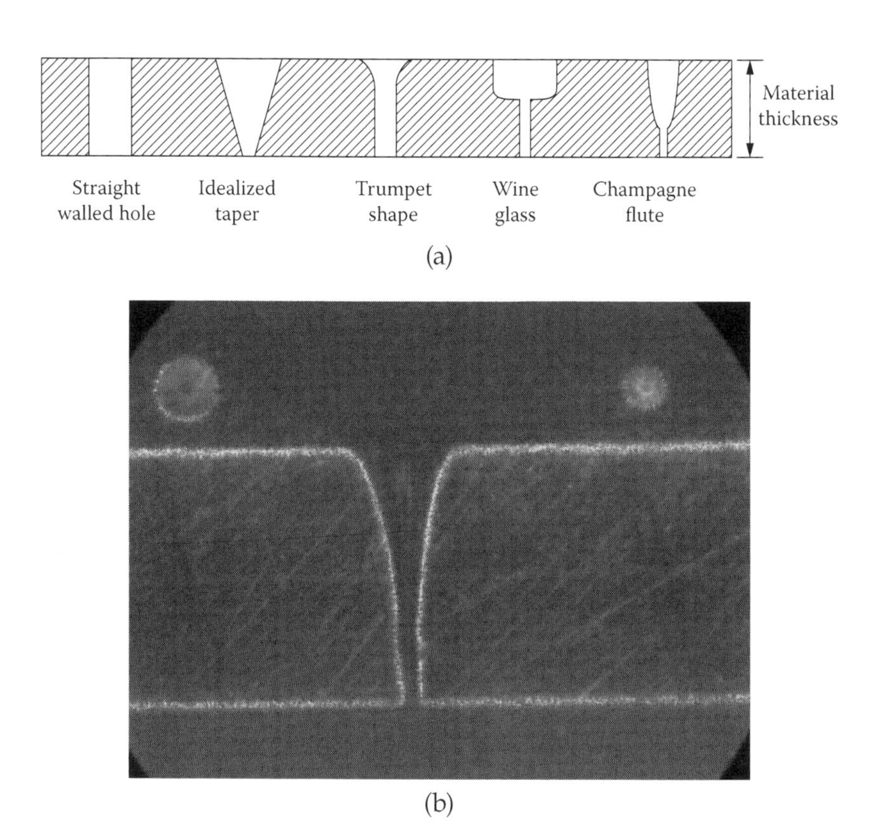

FIGURE 6.4
Cross sections of laser drilled holes: (a) idealized drawings; (b) "trumpet" shape in polyimide.

Figure 6.4(a) shows five different potential cross sections of a laser drilled hole. The straight walled hole (or minimal taper hole) is very difficult to realize in practice, but, fortunately, this hole shape is not needed except for some special applications. In fact, in most small-hole drilling applications, some structure in the hole profile is desirable. The idealized taper is one in which the entrance and exit holes can be measured and the taper is a simple linear function. In fact, what is normally seen when simply drilling a hole more closely resembles the trumpet shape as shown in both the drawing and a cross section in Figure 6.4(b), drilled in polyimide and cross sectioned. Finally, the wine glass and champagne flute profiles are useful for liquid or gas flow applications.

When small holes are drilled, especially in thicker materials (i.e., high aspect ratio), material content and thickness uniformity are crucial. Normally, taper is taken advantage of in order to get a very small hole diameter. This hole diameter is defined as the *smallest* diameter (D) of the cone, which is almost always the *exit* side. If we take into account the natural hole taper, we can drill holes much smaller than predicted mathematically, but it is very difficult to get good hole-to-hole uniformity in any case—especially if the material thickness or makeup varies. Counting the number of pulses, as one would do on a true drill through operation, may not work.

In these cases, some end-point detection can be used. Sensitive UV detectors can be used, sometimes coupled with light-gathering devices like an integrating sphere, to look at the bottom of the material and detect UV light breakthrough. Once breakthrough is detected, it is possible to count pulses while looking at the intensity of the breakthrough light and to stop within a pulse or two, even when drilling at 100 kHz rep rate. This technique has been used to drill holes of less than 0.5 μm in diameter with hole-to-hole variability of about 0.1 μm using the 355 nm laser. This can also be done at a very high speed by using the appropriate optics; for instance, the ability to drill over 150 parts per minute on a single tool with about 500 holes per part, with all holes less than 0.5 μm in diameter and all uniform to within a few percent, has been demonstrated.

The total time stated includes web motion, settling, and camera alignment as well as drill time, so laser-only drill time is much faster. The resulting profile resembles the wine glass in that the first burst of pulses is at higher energy and the final drill is done at greatly reduced energy per pulse. Figure 6.5 shows a front view of a roll-to-roll laser tool using two galvanometers (to increase throughput) and integrating spheres for end-point detection on each of the two beam lines.

The excimer laser can also be used and, in this case, if the holes are very small and very tightly packed in a uniform array, many holes can be drilled simultaneously. Thin plastic material has been drilled using a mask where

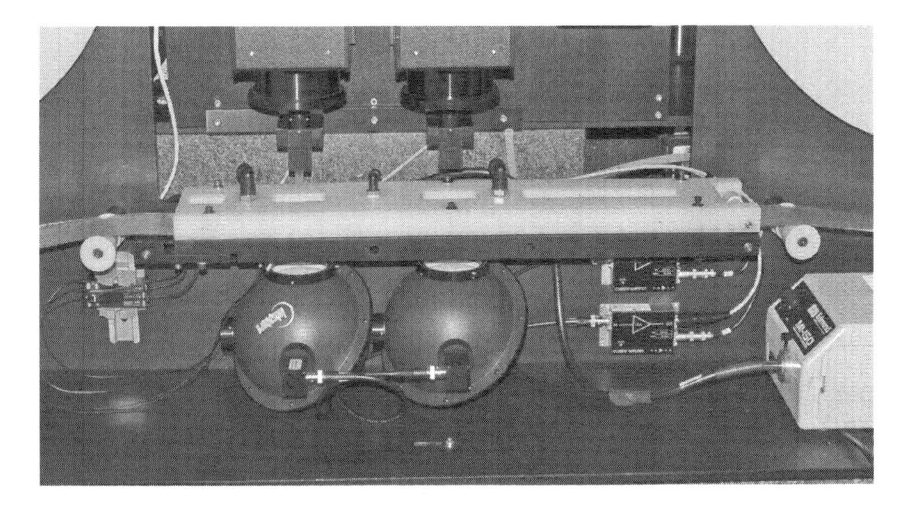

FIGURE 6.5
End-point detection using integrating spheres.

over 7,000 holes have been drilled simultaneously in one mask image. Wine glass and champagne flute profiles can be obtained by shuttling different masks into and out of the beam and/or by controlling the laser output energy. It is pretty straightforward to get holes less than 10 μm diameter using an excimer laser, assuming a reasonable aspect ratio—say, 20:1 or less. Using the preceding technique, holes have been drilled below 1 μm diameter. The downside of drilling many holes simultaneously is that end-point detection cannot be used for individual holes, but rather is averaged over the number of holes being drilled.

Here are a few of the applications requiring very small holes:

- *Interconnects* (satellites, medical devices, missiles and unmanned aerial vehicles, and cell phones). Here is an interesting tidbit: More people use cell phones on a daily basis throughout the world than wear shoes! Think about it. In fact, the technology exists to make cell phones much smaller than they currently are, but the typical human body is not equipped to deal with anything much smaller. There are thousands of lasers throughout the world drilling vias for electrical current conduction in cell phones alone.
- *Probe cards.* The holes drilled in these substrates hold an electrical conductor that is used to test wafers (normally a spring or wire); in this case, the taper is required to be as small as possible in order to hold the probe firmly. Current probe wire diameter is big by the preceding definition—around 4 mils (100 μm) or so—but forward looking research is looking at much smaller vias and, in fact, other technologies fail to reach very small hole diameters because of their inherent drawbacks (such as cleaning, plating, etc.). Figure 6.6 shows an assembled probe card with laser drilled holes in alumina ceramic.

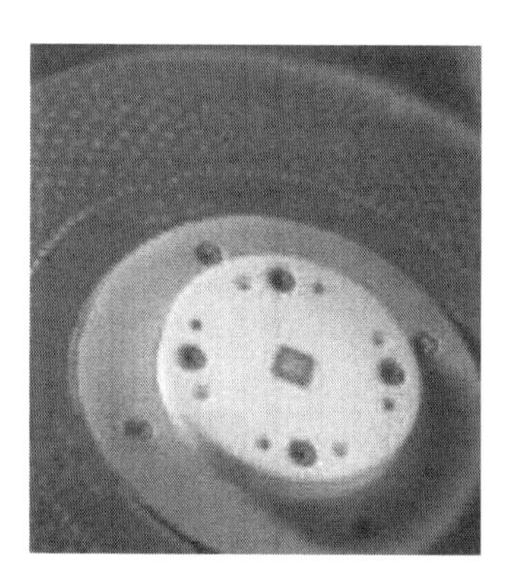

FIGURE 6.6
Assembled probe card made from alumina ceramic.

- *Filters* (blood cells, other cells, and organisms). Submicron holes are used to filter different biological species; these filters are used in cancer research, drug delivery, and manufacturing.
- *Drug delivery devices* (local delivery of drugs through plastic tubing or angioplasty balloons). Small holes are needed to control drug delivery. In the case of drilling directly into the balloon, the holes must be large enough to deliver the drug, but small enough so that the balloon can still be inflated. This can be done by drilling a matrix of very small holes that have an effective overall volume or open area.
- *Spray mist nozzles for analytical devices.* Many analytical devices, such as atomic absorption (AA) and inductively coupled argon plasma (ICAP) spectrometers require very small holes to form a fine mist that is looked at using either atomic absorption or emission lines. Sample preparation and presentation to the light are key to getting consistent and correct results.

In conclusion, UV lasers can be used to drill very small holes of less than 10 μm in diameter. If effective end-point monitoring techniques are used, even holes below 1 μm in diameter can be drilled quickly and reproducibly with minimal hole-to-hole variation.

6.3 Gas Assist

The industrial laser has been used since the early 1970s to process materials in a manufacturing environment, primarily for cutting, drilling, and welding. The earliest work was done on metals because early lasers were primarily infrared lasers that were more useful for metals than plastics and ceramics. Early materials included steel, stainless steel, and aluminum, but now include many alloys of, for instance, Cu, Ni, and Ti. Applications have subsequently been developed for cutting plastics, paper, alumina, and other ceramics, cloth, etc. Assist gases are critical to many of these processes and, in fact, many processes simply do not work without assist gas. Other processes may work without it, but the process is enhanced or sped up, or the quality is better with assist gases.

There are several reasons to use assist gases. The first is for active debris removal. Even UV lasers that break chemical bonds produce by-products, and efficient removal of these products from the processing area is beneficial. Usually simultaneous with this is the cooling effect that an assist gas can have on the process area. The next reason is to inhibit oxidation. Frequently, a cover gas is used to blanket the process area (this is sometimes done in an enclosed environment like a glove box) with an inert gas in order to prevent oxidation from occurring at the high temperatures associated with welding and cutting metals. Finally, the assist gas is sometimes used to *enhance* oxidation or some other process; in this case, the gas becomes a reactive component of the mix.

Assist gases are generally of more value with infrared (IR) lasers than with UV lasers. First, welding is generally not done at all with UV lasers, so this eliminates one very large application from the start. The short wavelength and associated short pulse lengths of the UV lasers do not lend themselves to joining applications. Second, for cutting and drilling with UV lasers, the use of assist gases generally has little or no noticeable effect on the part. There are several reasons for this. The short wavelength induces a different material removal mechanism. Also, UV spot sizes are generally much smaller than IR spot sizes, so there is less material to remove. Finally, with the exception of excimer lasers, UV lasers are generally delivered through galvanometer beam delivery systems, and these do not lend themselves well to coaxial gas processing.

Two areas where assist gases are used during UV processing are in pulsed laser deposition and laser cleaning of surfaces. Thin films of high-temperature superconductors (HtSc) can be deposited uniformly and rapidly using UV lasers and the appropriate gas environment. Other coatings, such as the deposition of thin films of crystalline and amorphous Si and of Cu by UV laser evaporation, can be applied by this method as well.

Laser cleaning systems for the semiconductor market remove organics from wafers. For instance, one company's patented cleaning technology uses "green" gases coupled with laser light to remove organic materials from the surface of silicon wafers and other substrates. The light and gas reaction creates a gas reaction zone (GRZ) where photochemical and/or photoablative removal of the selected material from the surface occurs. No solvents or chemicals are used during the process and only green inert gasses such as oxygen and nitrogen are used. For certain applications requiring an oxidizing reaction, oxygen is converted to ozone via a built-in ozone generator. The reaction does not require any additional heating and any heat generated by the reaction is localized to the reaction site and causes no damage to the underlying surface.

Several common gases are used in traditional laser processing. The first and most common is CDA (clean, dry air). Whatever gas is used, it is important for it to be pure and clean and dry. The presence of water will kill many processes, so using clean and dry air is important. Other gases usually have some purity specification and this should be considered when choosing a gas supplier.

For enhancing oxidation, an obvious choice is pure oxygen. This is used for cutting and drilling some metals and can generally be placed into the category of a reactive gas. In order to limit oxidation, as in welding of metals, an inert cover gas is used. This gas can be nitrogen, helium, neon, or argon. Lighter gases, like helium, move fast and can get into very small spaces easily, while larger gas molecules, like neon and argon, are very heavy and tend to be a better blanket. They also have enough mass—and therefore momentum—to deflect ejected material from the processing area. Sometimes even H_2 and CO_2 are used. Interestingly, it has been shown that sometimes gas mixtures perform better than pure gases.

Laser cutting and drilling in metals usually is accompanied by a HAZ (heat-affected zone) that sometimes requires a secondary process to remove. This is undesirable and can be minimized or eliminated using the right gas mixtures. For instance, steel responds well to a mixture of oxygen and nitrogen. Stainless steel and aluminum respond well to a mixture of N_2 and CDA, while Ti and Ni alloys respond well to Ar and He. Welding requires the assist gas to perform three main functions: protect HAZ from oxidation, minimize plasma effects in the weld area, and expel plasma from the weld joint. Helium has been the gas of choice because of its high ionizing potential and minimal metallurgical concerns, but it is expensive. Argon is less expensive, but it has a lower ionizing potential and the performance is not as good as that of helium. Here again, a gas blend may be preferable.

A few other things are very important. Gas pressure is extremely important and can vary greatly. For micromachining purposes, 80 psi is probably enough to do the job at hand, but in applications involving larger lasers, sometimes up to 300 psi is used with flow rates up to 1,000 standard cubic feet per hour. Finding the focus can be critical, and maintaining consistent results depends on maintaining the relationship between the work surface and the focal point of the beam. Beam centering in a coaxial delivery system is also extremely important, especially when consistent results are expected in multidirectional cutting. For the oxygen-assisted cutting of steel, for example, the concentricity should be within 50 μm. Nozzle centering is done after focus.

Finally, when coaxial gas-assist delivery is not possible (e.g., when using galvanometers), external nozzles can be used to direct the gas at the target area obliquely. The directionality of this gas can make a big difference in the quality on target—for example, if the gas is directed in the direction of or away from the direction of or perpendicular to the direction of travel.

Gas assist is very helpful in laser processing applications including cutting, drilling, welding, deposition, and surface alteration. It utilizes the gas pressure to assist in cutting and drilling and provides cooling of laser-generated heat and slag removal. Inert gases can be used as shields, and reactive gases can be used for cutting, cleaning, and deposition. Using quality gases from a known supplier is key to maintaining good processing conditions over a long period of time.

6.4 Micromarking

Three of the most important goals in advanced manufacturing are *miniaturization, compliance,* and *traceability.* Lasers are used as a tool for achieving these (and other) goals. In particular, laser marking is fast, cost effective, and indelible and can generate marks smaller than other competitive technologies. The flexibility of lasers allows them to be used easily on a wide variety of materials and, with the proper software, it is possible to generate marks in alphanumeric, bar code, or matrix code in static conditions and even on the fly. Marking can be achieved by etching the surface and removing material or by changing the surface structure; thus, at times it is not strictly machining, but because it is a huge market for lasers, some mention should be made. This discussion is confined to high-value micromarking applications.

Typical marked materials are metals, ceramics, and plastics. Metals are usually marked with an IR laser such as fiber or Nd:YAG. The etch depth is achieved by controlling the laser ON time or the number of passes. While it is possible to etch deeply into the material, micromarking usually occurs only on the surface of the part. For UV lasers, this mark can be on the order of fractions of a micron in depth, while for an IR laser, the depth will normally be much higher. The shorter the wavelength of the light, the smaller the minimum achievable spot size will be; for instance, it is easy to focus a 355 nm laser down to a 10 μm spot within a 25 mm field. Normally, seven distinct spots are necessary as a minimum for coding data (meaning that a 70 μm mark is possible in a relatively straightforward fashion). Smaller is possible using a UV laser and the right optics. UV light also interacts very well with many materials such as ceramics and plastics without "burning" the surface.

Laser marking is already cost effective in a host of applications where there is competition from traditional marking methods like etching and printing. However, as miniaturization proceeds in markets such as medical devices, microelectronics, aerospace, and defense, many traditional methods are no longer viable and lasers become not a complementary or competitive technology, but rather the only viable option for high-resolution marking of small parts. In the medical industry, every component that goes into medical implants will need to be traceable to its origin. This is also true of most aircraft/aerospace components as well as a considerable number of microelectronic components.

The equipment needed for marking is pretty basic. First, there is the lightbulb (LASER). The laser is chosen based on its wavelength, output power, size, and cost. Since in most cases marking does *not* in any way positively (or negatively) affect the actual operation of a component, any associated manufacturing cost is purely for cosmetics, traceability, or compliance. Therefore, this added cost must be low to keep the cost of the entire part at an acceptable level. Of course, in some applications—such as graduating an insertion catheter—the marks are a vital portion of the entire part. These applications can generally absorb a much higher per-mark cost. Low-cost threshold applications would probably need to use an IR laser, while higher cost applications might be able to justify the use of higher cost UV lasers.

Next, a beam delivery system is needed. Galvanometer-based optical systems are preferred since they can be easily programmed and they are fast. Galvanometers can travel at speeds of meters per second with accuracies of less than 25 μm. Field size is determined by the final f-theta objective, but is usually on the order of a few inches or more. The final necessary component is control software to drive the galvanometers and provide a user interface. Many versions of software are commercially available. The only additional things are a base/enclosure and perhaps some sort of holding fixture or part handling equipment.

Look at a typical marking job: Figure 6.7 shows a screen shot of the pattern to be marked and includes four types: alphanumeric, bar code, dot matrix, and a logo. First, the pattern is programmed into the software. Next, the height of the work piece is set such that the focus of the f-theta lens is on the surface of the part to be marked. After sizing, etc., the mark is made on the part. Surface marks require less energy and less dwell time than engraved marks, but in principle the process is the same.

Figure 6.8 shows the marks made on anodized aluminum. This material marks very well with almost any laser type and the anodizing only is removed, giving a very shallow surface mark with high contrast. Figure 6.9 shows the same mark made in white plastic using a 355 nm UV laser. Note that the UV mark is very crisp and clean, but still has high contrast and is indelible. Figure 6.10 shows the same mark in the same material using a 1 μm wavelength fiber laser. This IR mark, on the other hand, is burned into the material and has not only depth but also texture. This illustrates why a UV laser is used, for instance, for marking catheters. Not only is the mark barely legible, but also the roughness is a problem for inserting into the body.

FIGURE 6.7
Screen shot of mark file.

FIGURE 6.8
Mark of screen shot on anodized Al.

FIGURE 6.9
355 nm mark on white plastic.

Laser marking is already a well-established technique and this market is growing. Simple IR laser markers are portable, can be bought for less than $20,000, and are found at many diverse venues, including pet stores (marking custom pet-ID tags), biker rallies, and county fairs (etching images or photographs). There are almost 200 companies offering some type of laser marker and/or service just in the United States (not counting the retail outlets discussed previously). This means that the market is huge, but also that

FIGURE 6.10
1 μm fiber laser mark on white plastic.

competition in the marking service and system sectors is stiff. Since most consumer marking applications do not require high-end equipment and the software and optics are "off the shelf," it is very easy for anyone to obtain the equipment for most IR marking applications. UV marking requires a greater investment, but, even so, an enclosed 355 nm marker can be bought for less than $50,000. Therefore, laser markers will continue to play an ever more important role in product identification applications as miniaturization trends continue and the competitive technologies can no longer compete on a technical level.

6.5 Patterning Thin Films

One of the most important technologies involved in the miniaturization of many devices involves thin films. Depositing thin films is in itself a major technology area and in some cases lasers are used to vaporize a target material and deposit it uniformly over some surface. Most films, however, are generated using technologies other than lasers, such as chemical vapor or physical vapor deposition or spray coating. Because of the way these films are generated (uniform surface coated), if some areas should not be coated, a mask must be used in the deposition process or the films must be patterned after deposition. Using a mask has some inherent problems, including achievable resolution, potential damage to the part from physical contact, cost, and the need for inventorying masks of many designs. Other patterning methods include mechanical scribing and sandblasting. Both are problematic, with limited resolution, high potential to damage underlying substrate, and, in the case of sandblasting, dealing with the mess.

Fortunately, lasers can be used to remove films cleanly and precisely with very high resolution and competitive cost. The achievable resolution is much better than, for instance, lithography or mechanical processes—down to a few microns with a UV laser and short focal length lens. By selecting the appropriate wavelength and energy density on target, selective material removal can be achieved with little or no damage to the underlying substrate. These include optical films (dielectric masks or color patterns such as GOBOs [go before optics] for light projection and displays/show) and conductive films (miniature circuits, resistive heaters, cell isolation). Conductive films can be metal or metal oxides such as ITO (indium/tin/oxide).

What is meant by "thin"? Here, the definition of a thin film is that the film is less than 1 μm in total thickness. Most films are on the order of hundreds to maybe thousands of angstroms in thickness. This usually means that the films can be removed with one pulse of the laser because even UV lasers will remove 0.1 μm per pulse. It should also be noted that some films, notably metals, react quite differently in the bulk state than they do in the thin film state. For instance, clean removal of most metals in the bulk usually requires energy densities of 10 J/cm^2 or higher, while a thin film can be removed at almost the raw fluence of the laser—a couple hundred millijoules per square centimeter. However, many films do not even exist in the bulk.

Common substrates include metal, glass, ceramic, and plastics. These materials can be either completely opaque or transparent to some wavelengths; for example, glass is transparent to visible wavelengths of light. Transparency can be used to enhance processing; since the substrate is transparent, underlying damage from overexposure is minimized or eliminated. Also, frequently it is of interest to put the film side face down, or away from the incoming laser light, and direct the laser light through the substrate. This allows clean lift off of films and also helps keep debris to a minimum as it will tend to fall down and away from the glass rather than being redeposited on the substrate surface.

One example of patterning an optical film is used in displays and light shows. GOBOs are small (about 1 in. in diameter), transparent (in the visible range) substrates that are coated with light-reflecting films and then patterned using lasers so that, when projected, highly detailed and unique patterns are displayed. Edge quality is crucial since the projected image is many times larger than the GOBO itself. Also, because of the magnification, the smaller the lines or features, the more the resolution will be on target. Figure 6.11 shows a GOBO patterned using a 355 nm laser.

Another example is in the medical device field: disposable diabetes test strips. These test strips typically consist of several layers laminated together with windows for collecting the blood sample and chemistry between layers to initiate the reaction. One of the layers is a gold-coated plastic. The gold layer is a few

FIGURE 6.11
GOBO patterned using 355 nm laser.

hundred angstroms and can be easily removed at the low fluence of a 355 nm UV laser with little or no underlying damage to the plastic. The patterned gold forms the electrical circuit necessary for the digital readout of blood sugar levels. Figure 6.12 shows 25 μm wide lines patterned on a gold-coated Mylar™ test strip. This market is huge, and large numbers of these devices are needed at low enough cost so that the typical diabetic can afford to test several times per day.

A third example is huge for the laser market: photovoltaics. Lasers are used in the patterning of rigid and flexible thin film photovoltaics. Flexible substrates using conductive inks can be easily patterned with near-IR lasers

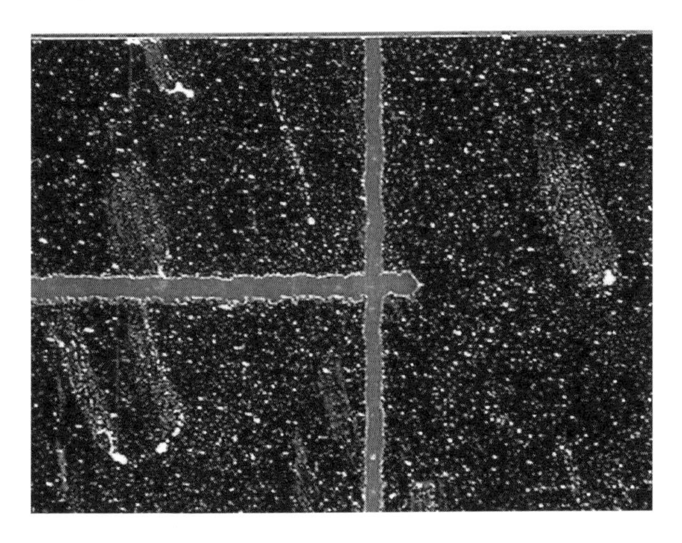

FIGURE 6.12
25 μm wide lines etched in gold-coated Mylar™.

on roll-to-roll processing lines, providing a continuous stream of product. While this technology has lower conversion efficiency than that of rigid products, it provides a great way to make low-power portable devices. Once the initial setup is paid for and in place, it is almost like printing money!

Rigid products—primarily Si-based thin film technology—use lasers to pattern the P1, P2, and P3 layers and also to do edge isolation. The P1 scribe is done back side by shooting through the glass with a 1064 nm laser to remove the transparent conductive oxide (TCO) and to define cells. The P2 scribe is then done after deposition of the Si using a 532 nm laser. The P3 scribe is then performed after deposition of the metal electrode, using a 532 nm laser that is transparent to the glass substrate and the TCO, but ablates the Si and metal. Edge deletion is performed in order to remove any thin films from the substrate edges so that a water and oxygen tight bond can be made between the glass sheets as these two are killers of conversion efficiency. Photovoltaic applications are described in more detail in Section 7.4.5 in the next chapter.

In the case of thin film patterning, lasers are the cleanest, most precise, and most cost effective manufacturing alternative. Sometimes, they are the only option available.

6.6 Multiple Hole Drilling Using Galvos

Galvanometers are typically used in micromachining applications with one incoming laser beam directed into the center of the input aperture and the beam expanded to fill the aperture to get the tightest focus on target. If the incoming laser beam is normal to the input aperture, the resulting spot on target will be centered in the focusing lens and therefore in the center of the field when the galvanometers are at their (0,0) position. This laser beam can then be translated over the field, and features can be made within that field by moving the galvanometer mirrors. Many applications require a large number of holes in a regular array, and drilling these holes one by one with a single laser beam requires a considerable amount of time. A certain technique uses multiple lasers and a single galvanometer to drill multiple holes simultaneously if the holes are in a regular array. Furthermore, it is also possible to make nonround features and to cover the surface (for area removal applications) by using a fill function.

It is possible to etch four holes simultaneously in an on-target substrate. A single fiber laser and a single galvanometer scanner can be used to demonstrate the principle. A fiber laser is easy to work with since it is small and portable, but in principle this technique can be used with any laser beam. Instead of aligning the laser beam centered on and perpendicular to the scanner's input aperture, the laser beam is aligned centered on and at an angle to the normal of the input aperture. The hole that is etched by the laser beam entering the input aperture at an angle is translated relative to where the hole would be if the beam was perpendicular to the input aperture. By properly selecting the four input angles and sequentially positioning the single fiber laser to create those four input angles, four holes can be etched in a square pattern with any center-to-center spacing that fits within the galvanometer field. In fact, if four holes are made at the corner of a large square, it can be seen that in principle at least nine laser beams or even more can be used simultaneously if the holes are closely spaced and physical room is sufficient to fit the lasers and beam expansion telescopes.

First, a single laser setup is used to determine the relationship between laser beam input angle and hole spacing, using a single fiber laser and a 30 mm aperture scan head with f = 163 mm f-theta lens. The needed incoming angles can be calculated and an alignment mask can be generated that simplifies the setup, as shown in Figure 6.13. The alignment mask is placed on alignment pins at the input aperture entrance. Alignment holes, used to align the lasers, have a diameter of 1.5 mm and come in sets of four that form the corners of a square. The centers of each set of four alignment holes lie on an alignment circle of known diameter.

The actual setup uses multiple lasers to etch multiple holes simultaneously. For example, consider using four lasers and drilling four holes simultaneously. A detailed view of the lasers and optics is shown in Figure 6.14. The four lasers are located so that their beams will pass through the center of the scanner's input aperture and at angles that will produce an array

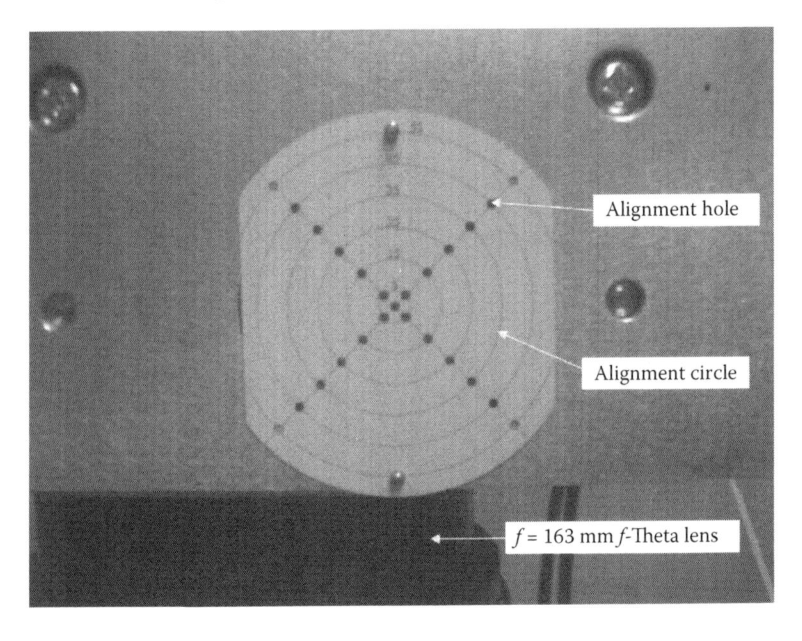

FIGURE 6.13
Alignment mask.

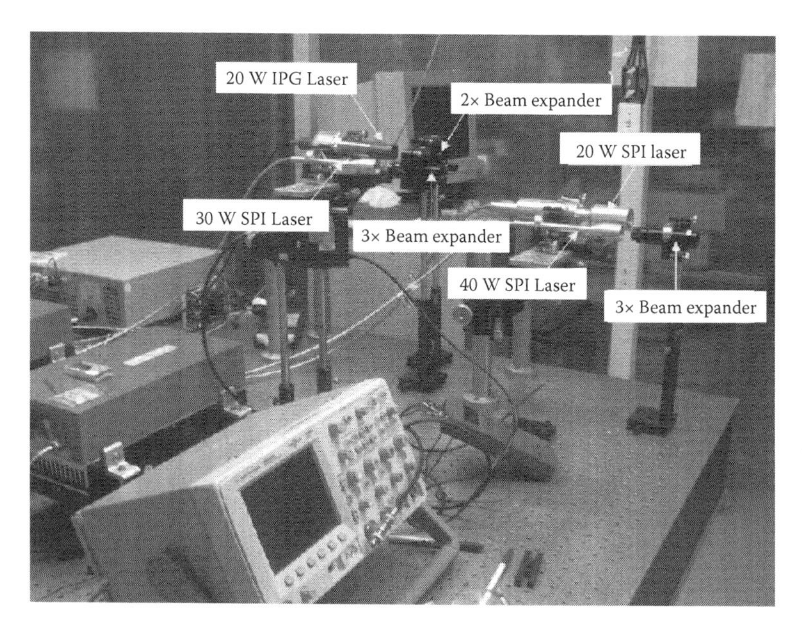

FIGURE 6.14
Expanded view of the lasers and telescopes.

of four holes at the corners of a square. Note that the use of four identical lasers greatly simplifies the setup because different lasers will have different divergence and raw beam sizes that must be accounted for. Long distances between the lasers/beam expanders and the input aperture of the scanner are needed to accommodate the equipment physically and provide clear lines of sight.

A photograph of the four holes being etched simultaneously is shown in Figure 6.15. On the photograph, the laser that is etching each hole is

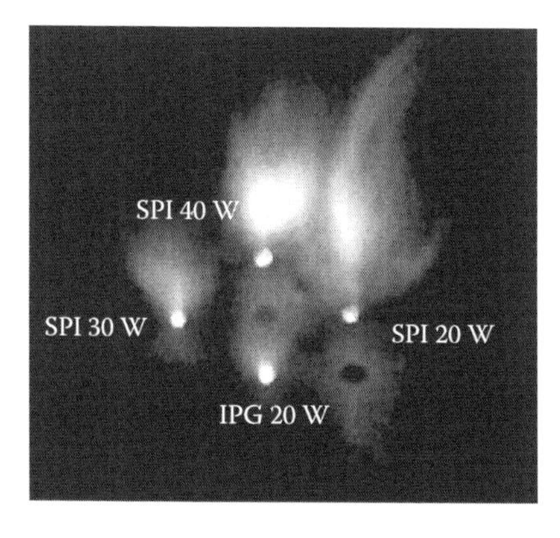

FIGURE 6.15
Four holes drilled simultaneously.

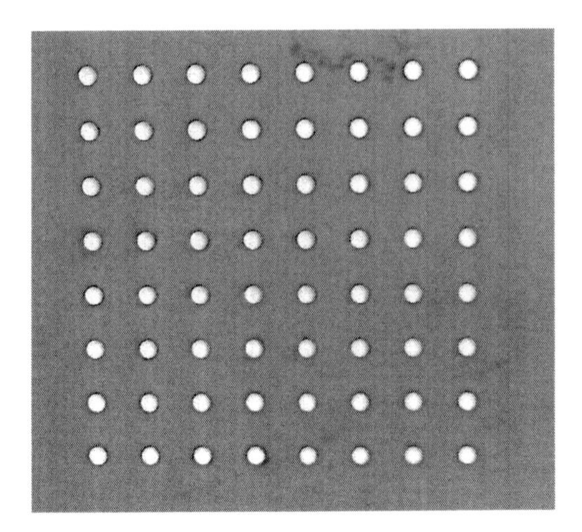

FIGURE 6.16
8 × 8 array of holes etched in a flat substrate.

identified. The image was acquired from above the sample, looking from the third quadrant toward the first quadrant relative to the X, Y scan axes; this explains the fact that the holes do not appear to be in a square array.

Figure 6.16 is a photograph, taken with back lighting, of an 8 × 8 array that was etched using the procedures described before and the equipment shown in previous figures. The time to etch the 64 holes shown in Figure 6.16 depends on the material type and thickness and is limited by the laser power, but in one case 1 mm diameter holes were etched in 1 mm thick material in about 20 seconds. The holes were etched four at a time but the four holes being simultaneously etched were not next nearest neighbors; this helps reduce the heat load and also reduces the space needed for the optical setup since the input angle is greater for holes spaced further apart.

Another reason that fiber lasers were chosen is that, on a dollars-per-watt basis, they are among the most inexpensive photon sources around. A 20 W laser currently sells for less than $15,000, and it is easy to imagine using up to nine of these lasers with a total laser cost of less than $135,000—rather than, for instance, a high-power UV laser or ultrashort-pulse laser with an individual laser price of $150,000 to $300,000.

This technique only applies to surfaces that are flat, unless the radius of curvature is minimal; however, in concept, dynamic control of the beam expansion telescope can be used to change the focal distance. The technique applies not only to round hole drilling but also to various other shapes that can be accommodated. This makes this method more flexible than using rotating prisms, for instance. Also, by using a "fill function," large areas of material can be removed in applications such as removal of thermal barrier coatings or paint from aircraft components. If the surface topography is known or if clever in-line feedback is used, each individual laser can also be turned off and on dynamically.

6.7 In-Volume Selective Laser Etching (ISLE)

Hard and brittle materials such as different types of glass and sapphire are difficult to work with, even using laser technology. Poor absorption at most wavelengths combined with poor heat transfer properties makes the realization of structures in these materials difficult or impossible. A two-step technique for high-resolution structuring using short femtosecond laser pulses has been developed and demonstrated by the team "In-Volume Structuring" at the Fraunhofer Institute for Laser Technology (ILT) in Aachen, Germany. This technique uses focused laser pulses to change the chemical structure of the material such that subsequent immersion in a proper wet etching agent (depending on the material) selectively etches only the exposed area, leaving the unexposed area untreated. This technique is used for surface and in-volume structuring and has potential applications in the areas of microfluidics and the manufacture of three-dimensional parts.

The provided energy is absorbed only in the focal volume where the intensities are high enough for nonlinear absorption processes. This has the effect of changing the structural properties of the lattice, resulting in a density and refraction index change of the exposed area. The femtosecond laser pulse is used in conjunction with a microscope objective in order to get the required high intensities on target. The resulting spot size is about 1 μm. A high repetition rate laser must be used with a fast scanner in order to irradiate the volume with speeds up to $v > 0.3$ m/s. However, the repetition rate should be small enough to avoid heat accumulation effects.

After exposure, the sample is immersed in the wet etching agent to remove the exposed material selectively. For instance, the etching ratio between sapphire and the resulting amorphous sapphire is about 1:10,000. This method can thereby produce features with a surface roughness of Sa = 64 nm, which is approximately the value of the roughness of current highly polished mechanical devices like bearings. Furthermore, very small material loss is achieved; parts can be totally excised from the bulk with less than 2% total material loss. Consider the following examples:

In-volume marking. Micromarking in transparent materials is achieved *below* the polished surface. Figure 6.17 shows a field of squares in a depth of approximately 200 μm in fused silica. Each square was structured in less than 3 seconds using intersecting line segments with varied 3–10 μm pitch between lines. The modified lines exhibit a refractive index change with no observable cracks. Due to the diffraction from the micrgrid and interference in the observer's eye, colors are observed when varying the angle of viewing.

FIGURE 6.17
Colorful marking in the volume of fused silica; detail at right. (Courtesy of Fraunhofer ILT.)

Two-dimensional ISLE for cuts. Figure 6.18 shows a scanning electron microscope cross section of a microslit in sapphire. This slit was made by using 50 overlapping tracks while adjusting the z position between tracks to get the depth. The dimensions are 1 cm long (y) by 125 μm deep (z) by 1 μm width (x). Three-dimensional parts can also be manufactured by using this strategy, generating hollow free-form structures and channels for microfluidic devices. The dimensions of the structures are controlled by the number of adjacent tracks.

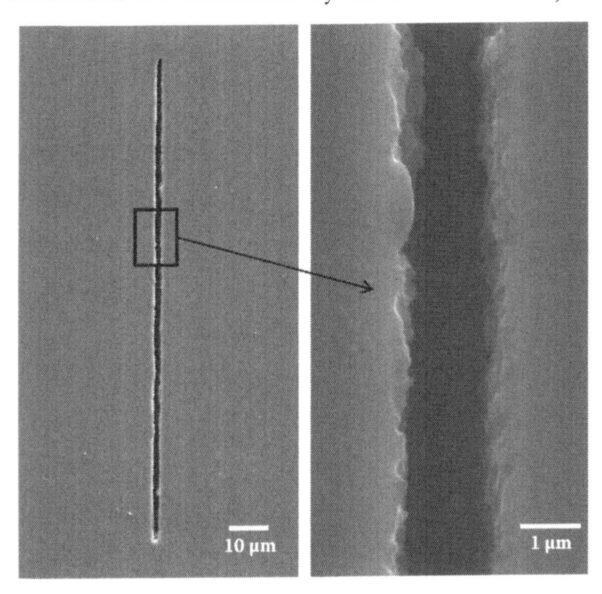

FIGURE 6.18
Cross section of a microslit in sapphire. (Courtesy of Fraunhofer ILT.)

Currently, channels are etched on two plates, which are then joined together. This technique creates many problems and is also a multistep process. With the ISLE technique, large hollow structures are created in a two-step process out of bulk material. Cubes of dimension 50 × 10 × 10 μm have been created. The obvious condition is that channels must be etched from the surface for the wet etching agent to flow into the proposed cavity.

Three-dimensional ISLE for manufacturing microparts and microholes. Figure 6.19 shows a cylinder removed from a bulk substrate by exposing the cylinder "wall" all the way through the sample and etching. The diameter is 500 μm and the part thickness is also 500 μm. After etching, the cylinder easily falls out of the bulk material. The cut width of about 1.4 μm is shown in Figure 6.20 before the resulting cube was extracted. The cut dimensions are about 450 × 500 × 500 μm. The volume of the kerf (450 × 2000 × 1.4 μm) compared to the volume of the cube shows a material loss of less than 2%. Figure 6.21 shows a gear that was extracted from bulk 1 mm thick fused silica. Structures 1 mm in height and 1 mm in diameter have been laser irradiated for 20–40 seconds.

This ISLE technique is very exciting and has only been recently developed by using short pulse femtosecond lasers and fast optics. The technique has some very powerful applications, especially in the medical field, and also for sensing and perhaps gas flow dynamics. Using existing CAD/CAM software enables the rapid manufacturing of microstructured parts and in-volume micromarkings by exploiting the potential of today's high-repetition rate femtosecond lasers. Crack-free modifications inside sapphire and fused silica have been produced and colorful marking has been written at high speeds. Commercial applications of this technique are in progress.

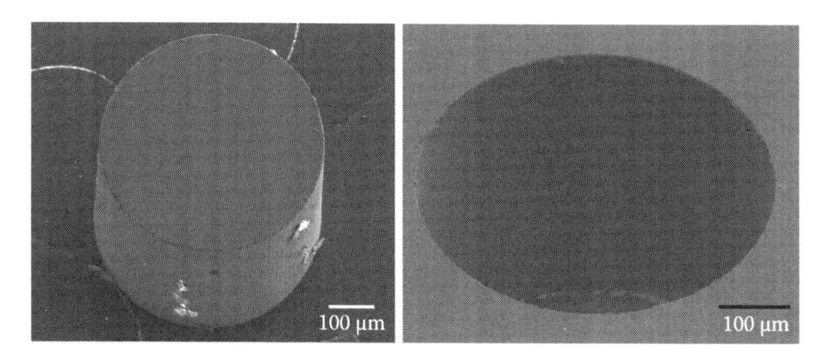

FIGURE 6.19
Cylinder with 500 μm diameter (left) and the substrate from which it was removed (right). (Courtesy of Fraunhofer ILT.)

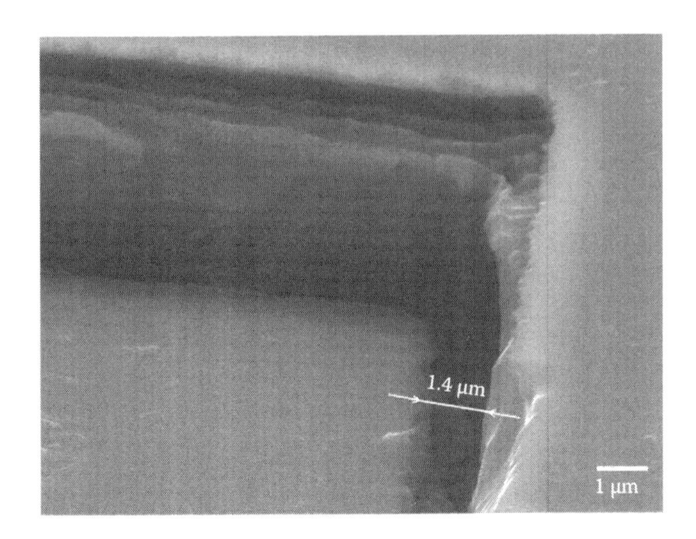

FIGURE 6.20
1.4 μm kerf before cube removal. (Courtesy of Fraunhofer ILT.)

FIGURE 6.21
Gears etched from 1 mm thick fused silica. (Courtesy of Fraunhofer ILT.)

7 Applications

7.1 Microelectronics and Semiconductors

These two areas are grouped together, although they might be thought of as different markets. Currently, most of the volume microelectronics applications are in Asia. The markets that remain in the United States are for the most part related to defense and security, quick turn and prototyping, or small volume that is not of interest to larger corporations.

7.1.1 Microvia Drilling

Microvia drilling in printed circuit boards and flexible circuits is a huge business with hundreds of lasers (mostly in Asia) drilling billions of microvias annually. A microvia is a small hole drilled for electrical conductivity in printed circuit boards and flex circuits. The diameter of these vias is 50–200 μm. Important advantages of laser processing include speed, smaller achievable hole size, cleanliness, hole placement accuracy, and cost-to-performance ratio. Hole drilling speeds depend on material type, thickness, laser type and power, and hole pattern; thousands of holes per second can be drilled under the right conditions.

High-density hole drilling enables portable electronics manufacturing as more precision and efficiency allow for more design options. With the use of thinner materials, ultraviolet (UV) lasers (typically 355 nm) can be used for both the copper and dielectric, although great care must be taken to avoid etching through the underlying copper conductor. A better way is to etch the top copper layer chemically or to drill the surface holes with a 355 nm UV laser and then use a CO_2 laser to remove the dielectric down to the underlying conducting layer. This has the virtue of removing the dielectric more rapidly and also the underlying copper acts as a barrier since copper is highly reflective of the 10 μm wavelength. This only works for vias with diameter greater than about 75 μm, below which a UV process must be used.

Figure 7.1 shows some plated laser-drilled microvias in FR4 (flame-resistant Class 4) material. This material is somewhat harder to work with than others in that it is inhomogeneous and the embedded fiber glass ablates at a

FIGURE 7.1
Plated laser-drilled microvias in FR4.

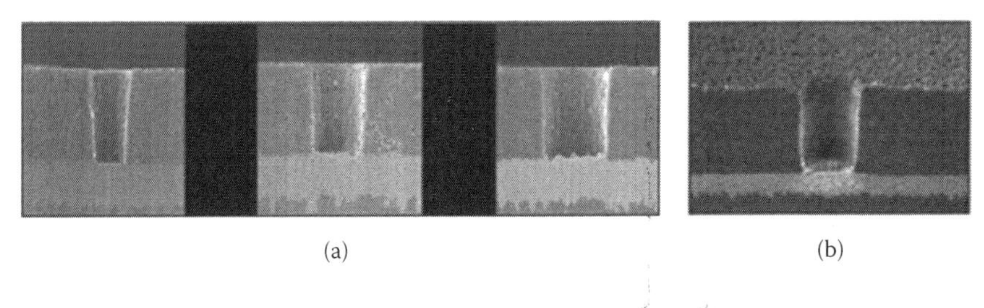

(a) (b)

FIGURE 7.2
(a) 30, 40, and 50 μm diameter vias in resin-coated copper; (b) oblique view. (Both images courtesy of Hitachi Via Mechanics, USA.)

much higher fluence than the epoxy matrix, but it is relatively inexpensive and is probably the most frequently used dielectric. Figure 7.2(a) shows 50, 75, and 100 μm diameter UV laser-drilled vias in resin-coated copper, while Figure 7.2(b) is taken at an oblique angle. Homogeneous materials work much better for UV lasers than glass-filled material since the glass is somewhat transparent at 355 nm. If done correctly, the UV light can clean the underlying copper such that no subsequent cleaning process is needed before plating, but in practice a cleaning step is usually performed. In fact, if short pulse UV light is used, 50 μm blind vias in circuit boards can be drilled at the rate of hundreds per second, with little or no heat-affected zone and in any common circuit board material.

7.1.2 Dielectric Removal from Conductive Surfaces

Another active area is in removing dielectric materials from conductive surfaces. The fastest method of accomplishing this is to use a CO_2 laser, which has the virtue of being almost completely reflective off copper; therefore, there is less chance to etch through underlying conductive layers. Because it is a thermal process, there is a HAZ (heat-affected zone) around the edges and there may be debris that needs to be cleaned afterward. More importantly, when the laser gets very close to the reflective copper surface, the long penetration of the 10 μm wavelength, coupled with less than 100% absorption, leaves a thin residue on the surface that needs to be cleaned subsequently. However, a cleaning process before plating is normal anyway. If a UV laser is used, the copper surface is in principle clean enough for plating afterward, but, again, a cleaning process is normal—especially if the subsequent plating does not take place immediately and oxidation occurs.

Flexible circuits are the most common products that have laser-ablated conductive pads. Figure 7.3 shows excimer laser removal of polyimide-coated copper using mask imaging to expose all the conductors simultaneously. This product was done on a roll-to-roll machine. Figure 7.4 shows conductors exposed using a CO_2–TEA laser.

Another opportunity exists in repairing engineering or manufacturing mistakes. For instance, solder mask is extremely hard to remove once it is applied (unless lasers are used, which makes the task pretty straightforward). Solder mask is put in the wrong places on a routine basis and there is cost associated with remanufacture—plus, by the time the mistake is noticed, delivery may be impacted. Therefore, lasers can be used to minimize rework costs and to achieve on-time delivery. Figure 7.5(a) shows a CO_2–TEA laser removing solder mask from a printed circuit board; Figure 7.5(b) shows the end result with the untouched section on the left and the laser-processed section on the right. Figure 7.6 shows the result when one tries to remove solder mask mechanically (left) and the same mask removed by a laser (right).

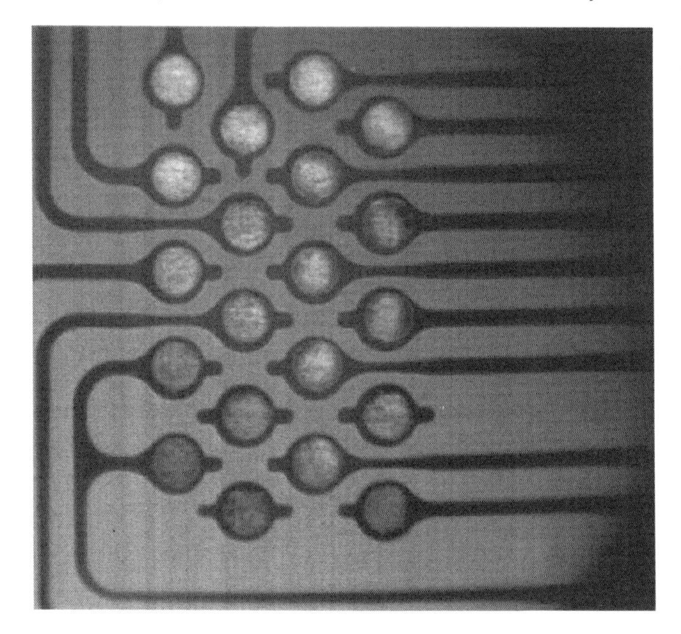

FIGURE 7.3
Excimer laser dielectric material removal.

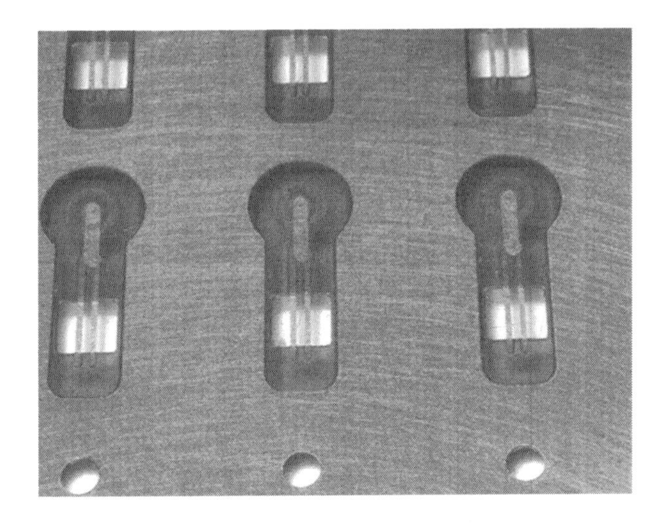

FIGURE 7.4
CO_2–TEA dielectric material removal.

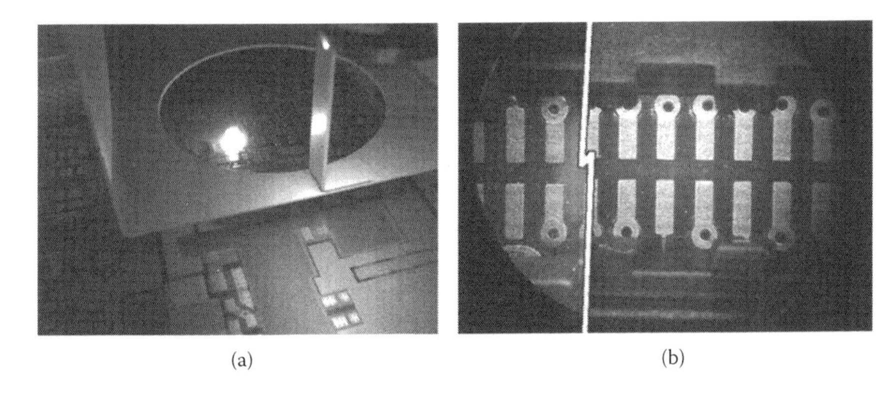

(a) (b)

FIGURE 7.5
(a) CO_2–TEA laser removing solder mask from a PCB; (b) laser-processed area on right.

7.1.3 Solder Mask Stencils

Lasers are the preferred method for generating the solder mask stencils that are used to lay down the mask. The preferred method is to use an infrared (IR) laser (YAG or fiber) to produce the multitude of features rapidly on a circuit board. Low taper features with excellent edge quality can be achieved using a high-power laser with minimal postprocessing needed. Figure 7.7 shows a stainless steel solder mask being made using an IR laser. Stainless steel masks in thicknesses from 20 to 600 μm can be made using IR lasers, but there are limitations in attainable feature size. When the smallest features are required, a UV laser can be used to pattern polymeric masks, as shown in Figure 7.8.

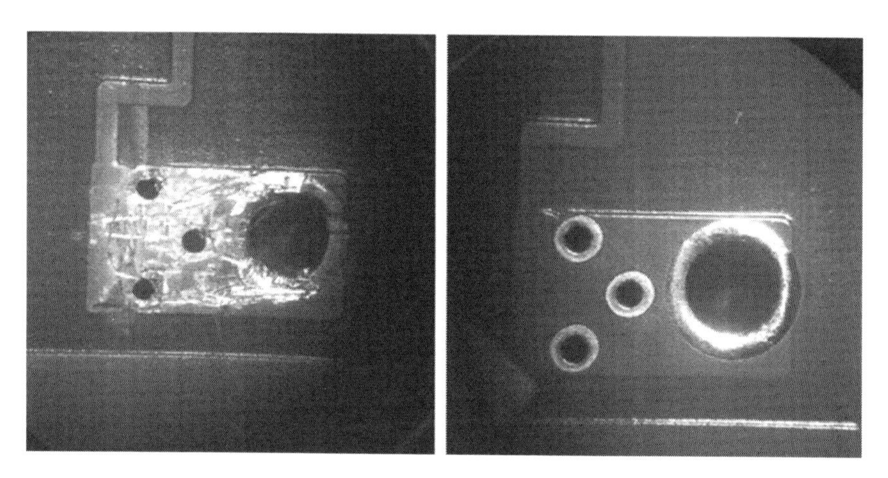

FIGURE 7.6
Mechanical (left) and laser (right) removal of solder mask.

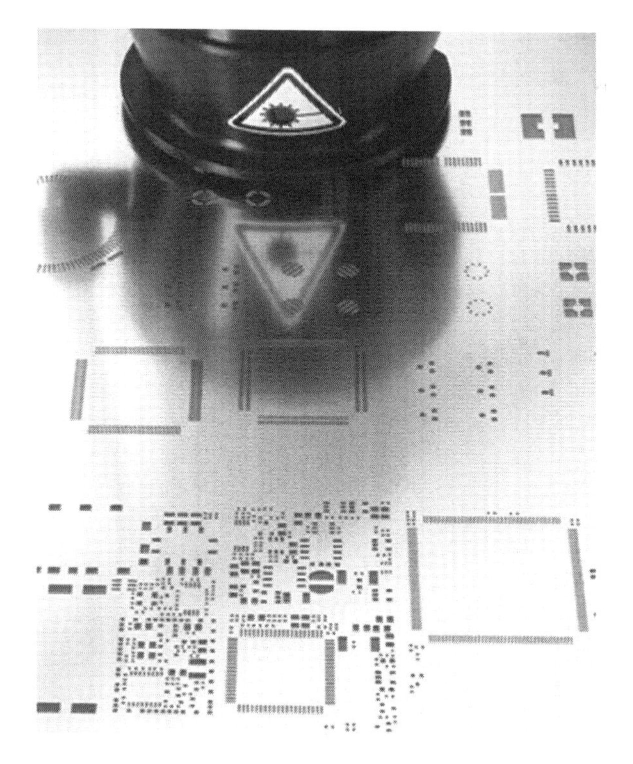

FIGURE 7.7
Stainless steel solder mask stencil produced with an IR laser. (Courtesy of LPKF.)

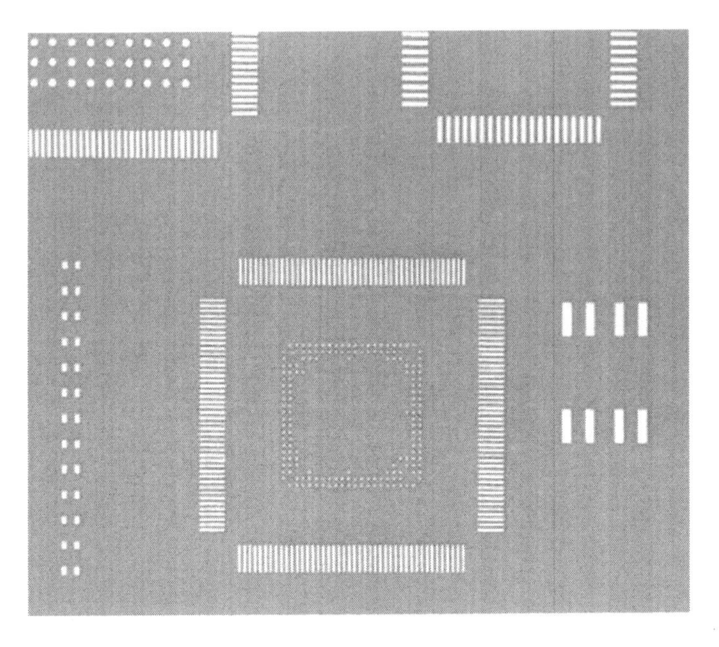

FIGURE 7.8
Polymeric solder mask stencil. (Courtesy of LPKF.)

7.1.4 Short Repair

During the manufacturing process, conductive material sometimes "bleeds," so shorts occur. If these shorts can be identified by optical or electrical testing, lasers can be used to isolate the conductors and make the part usable. This is usually done with a UV laser as it will cut the metal conductor while doing minimal damage to the underlying material or components. On a large, complicated device, sometimes hundreds of shorts cannot completely be eliminated in the standard manufacturing process; in this case, lasers can be used as a repair tool. Testing equipment can identify problem areas with precise coordinates and a laser repair tool can take these coordinates and automatically make repairs by cutting the shorting conductor. The laser tool is usually trained to recognize a finite set of frequently occurring faults and do the repairs automatically, but when a fault is found that is not in the listed database, an operator can be called to evaluate the situation and make a decision on how to address the problem.

7.1.5 Indium Tin Oxide and Conductive Metal Structuring

Indium tin oxide (ITO) is a highly conductive thin film. It is transparent and colorless in thin layers. ITO is one of the most widely used transparent conductive oxides (TCOs) because of its two chief properties: electrical conductivity and optical transparency. ITO can be easily deposited in a thin film using a range of evaporation or sputtering techniques. As with all transparent conducting films, increasing the thickness will increase the material's conductivity, but decrease its transparency. Typical film thickness is hundreds or perhaps thousands of angstroms.

Lasers work extremely well in the patterning of ITO films and such devices are used in flat panel displays, touch screens, instrument displays, and some medical devices. Lasers are used to isolate certain portions resulting in an electrical circuit. Green (532 nm) lasers are very interesting, especially for use on displays as the substrate is usually transparent in the visible portion of the spectrum. Therefore, no damage is done to the underlying substrate. If a large amount of material is to be removed—for instance, on a glass panel—it can be done by lasering through the glass with the ITO facedown, resulting in no generated debris being deposited on the surface of the part. For smaller spot sizes on target, 355 nm also works well. Figure 7.9 shows ITO removal from glass. The line width is about 25 μm; the patterning is invisible to the naked eye.

The same lasers can be used to structure thin metals to form conductive circuits. Metals thicker than a few microns exhibit the bulk properties of the metal, but at less than that the ablation properties change somewhat in that they can be easily removed at a much lower ablation threshold than in the bulk. Lasers can be used to structure thin metallic films as described before for ITO. This is useful in direct writing of microcircuits.

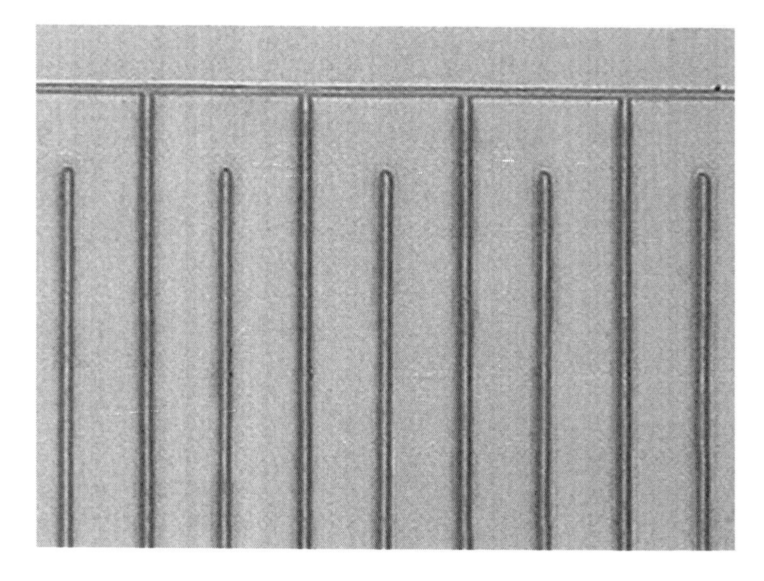

FIGURE 7.9
ITO removal from glass.

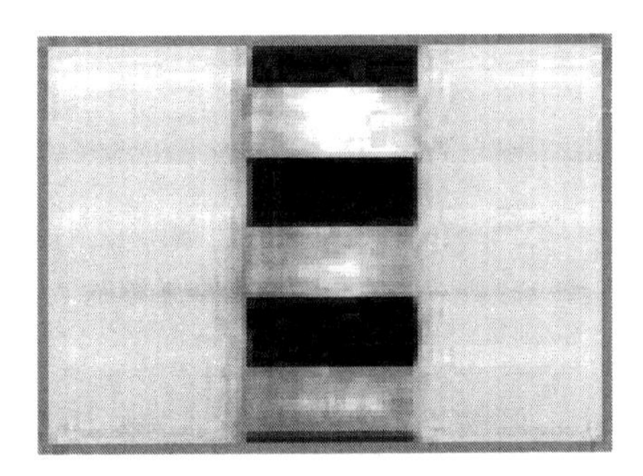

FIGURE 7.10
CO_2 laser wire stripping. (Courtesy of Synrad.)

7.1.6 Wire Stripping

Larger diameter wires can be "stripped" with a CO_2 laser in a two-step process. Figure 7.10 shows a set of conductors exposed on both sides using a 20 W CO_2 laser. First, the insulation is removed on one side, and then it is turned over and the process is repeated. Round wires can be cut around the circumference and then the insulator is manually removed. If a stripped section is in the middle of the wire rather than on the end, a longitudinal cut is made so that the insulation can be removed. If the wire and insulator are thin enough, lasers can be used to remove all the material completely without a secondary manual step. This is done using either a CO_2–TEA laser or an excimer laser.

Figure 7.11 shows an optical schematic for excimer laser wire stripping of thin (40 gauge, which is about 80 μm diameter) wire coated with either polyimide or polyurethane. In order to inhibit oxidation, the stripped ends are manually solder-tinned in a pot. Figure 7.12 shows (a) unprocessed, (b) laser stripped, and (c) tinned wire. Microflex circuits have taken over many of the consumer applications for stripped wires (for example, data storage devices), but there is still demand for use in sensitive electronic applications, magnet wire, and gyroscope wires.

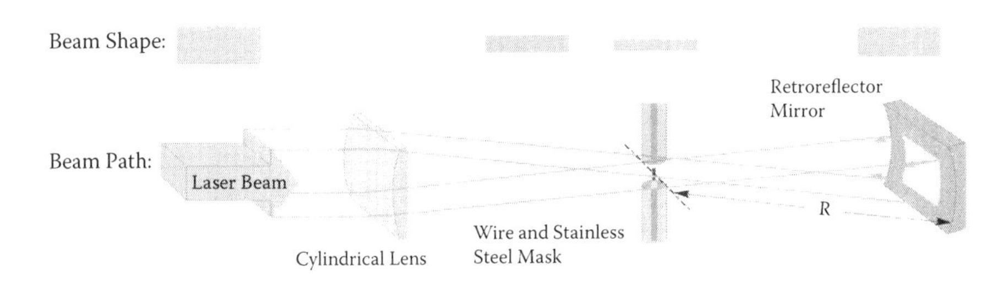

FIGURE 7.11
Optical schematic for wire stripping.

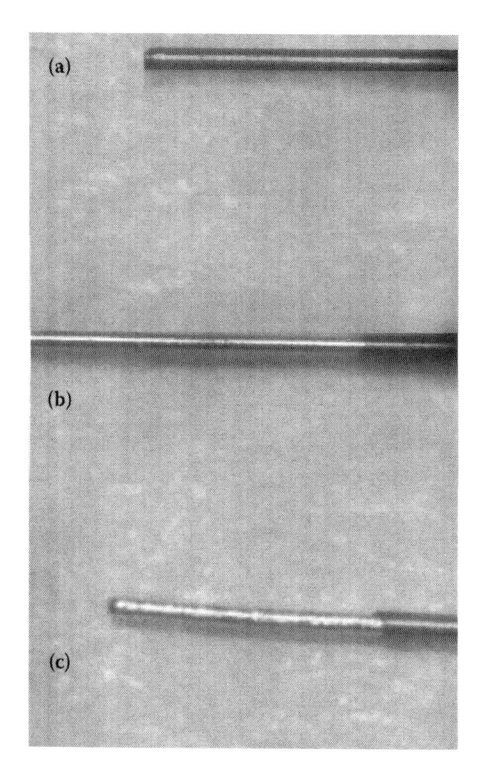

FIGURE 7.12
(a) Unprocessed, (b) stripped, and (c) tinned wires.

7.1.7 Resistor Trimming

Lasers can be used to adjust the operating parameters of an electronic circuit. The most common application uses a laser to ablate away small portions of resistors, raising their resistance value. The operation is usually conducted while the circuit is being simultaneously tested, leading to optimum final values for the resistor(s) in the circuit. During the trim process, the corresponding parameter is measured continuously and compared to the programmed nominal value. The laser stops automatically when the value reaches the nominal value. The resistance value of a film resistor is defined by its geometric dimensions (length, width, height) and the resistor material. A lateral cut in the resistor material by the laser narrows or lengthens the current flow path and increases the resistance value. The same effect is obtained whether the laser changes a thick film or a thin film resistor, regardless of substrate.

Capacitors can also be trimmed. The reason for using lasers is that it is not always possible to manufacture components operating at very specific values, so a laser trimming step is used in the manufacturing process. The actual trimming operation is quite straightforward, but the key is to use active feedback and do the trimming extremely fast in real time. A laser trimming system consists of a laser with high beam quality and small beam diameter (20–40 μm), an industrial PC with appropriate software, a galvanometer beam delivery, instrumentation to measure the circuit parameter, a probe to make contact with the circuit, and an X–Y table. Frequently used cut patterns are straight lines, L-shape, serpentine, and large area removal.

7.1.8 Radio Frequency Identification

Radio frequency identification (RFID) technology is transforming the way in which major corporations do business. RFID is a system that uses radio waves to transmit the identity of a person or object wirelessly. Unlike bar coding, RFIDs require no direct contact or line of sight for communication, and hundreds of tags can be read simultaneously and not serially as with bar coding. RFID can operate through the body, clothing, and a host of other nonmetallic substances, as well as at some distance. Common examples include EZ Pass for road tolls and gas station speed passes.

A typical RFID device consists of a microchip with up to 2 kilobytes of information attached to a radio antenna mounted onto a substrate. These devices are used in asset tracking, manufacturing, supply chain management, retailing, security and access control, and payment systems. The cost for passive devices (not battery operated) is about $0.05; more sophisticated devices go for several dollars or tens of dollars. IR and UV lasers are used for etching the antenna, fine-tuning the circuit, slitting, scoring, and drilling. Typical feature sizes are in the range of 50–100 μm.

7.1.9 Microelectromechanical Systems Components

*M*icro*e*lectro*m*echanical *s*ystems (MEMS) are very small mechanical devices driven by electricity. The size of the total device is generally between 20 and 1000 μm; each individual component is in the range of 1–100 μm. MEMS device structures can be fabricated by direct laser removal of material. Because many bioabsorbable or compound materials are not well suited to conventional fabrication techniques like chemical or plasma etching, laser machining is becoming a frequently used manufacturing technique. Lasers can be used directly on the devices or they can be used to make masters from which mass production is accomplished using replication techniques. Depending on the material and features, lasers from the IR through the UV can be used for MEMS devices.

In volume production, it is more likely that a variety of different techniques, including lasers and etching, are used. It is also possible to use a laser beam focused into the depth of the substrate, usually silicon, and scanned along the dice lanes. Gentle mechanical tension is generated by stretching the dicing tape to which the wafer has been mounted. This causes the material to fracture cleanly along the laser exposure path. The result is a clean break with no kerf loss or chipping.

7.2 Medical Devices

Currently, 50% of all the disposable medical devices used worldwide are made in the United States; about 50% of this manufacturing capacity is outsourced. This market is somewhat difficult to penetrate initially. Of all the projects that get started, only about 1 in 10 matures to full production. Also, there is typically a long lead time associated with product approval. However, on the positive side, lasers play an enormous role in the manufacturing of disposable medical devices and the market is large. Also, once a company is qualified as a vendor and volumes go up on specific products, that company is pretty much locked in as a vendor as long as deliveries, pricing, and quality standards are continuously met. Some of these product cycles can run for years.

7.2.1 Diabetes Test Strips

Millions of test strips per day are manufactured; a large number of the manufacturers use lasers to pattern an electrical circuit on gold-coated polymer. Blood is drawn into the test strip, where it reacts with enzymes, and then electrochemical technology is used to produce a reading. This reading is correlated to a glucose value—the number seen on the reader display. The strip manufacturing process uses multiple layers of materials. While all of these different layers can be made using lasers, only the base material, made of a thin gold coating on a rigid plastic, uses lasers in production because the other layers do not require the precision and can be stamped. UV lasers are used to etch electrodes into the gold conductor.

Figure 7.13 shows a thin gold film patterned on a clear polymer. This pattern is much more complex than a real test strip and it can be done about as fast as the beam can be moved over the part. Metallic coatings like this are removed cleanly in one pulse of the laser. Lasers have moved manufacturing of these devices to the point where only a small amount of blood is needed for an accurate measurement.

However, a better scenario is getting a blood glucose reading without pricking the body for blood. Figure 7.14 shows a ferrite plug machined with a 355 nm laser. Ferrite material can be used to make implantable devices that send a radio signal to a receiver (current devices look a bit like a wristwatch) that translates into a blood glucose level. This material is very difficult to pattern by any other method as microcracks or HAZ results in damaged devices during firing. No blood is required and readings can be taken as often as desired; in addition, all readings are digitized and can be stored or shared with a physician. Placement of the device in the body is an outpatient procedure taking only a few minutes. This technology, while not in general use, is being developed using both IR and UV lasers.

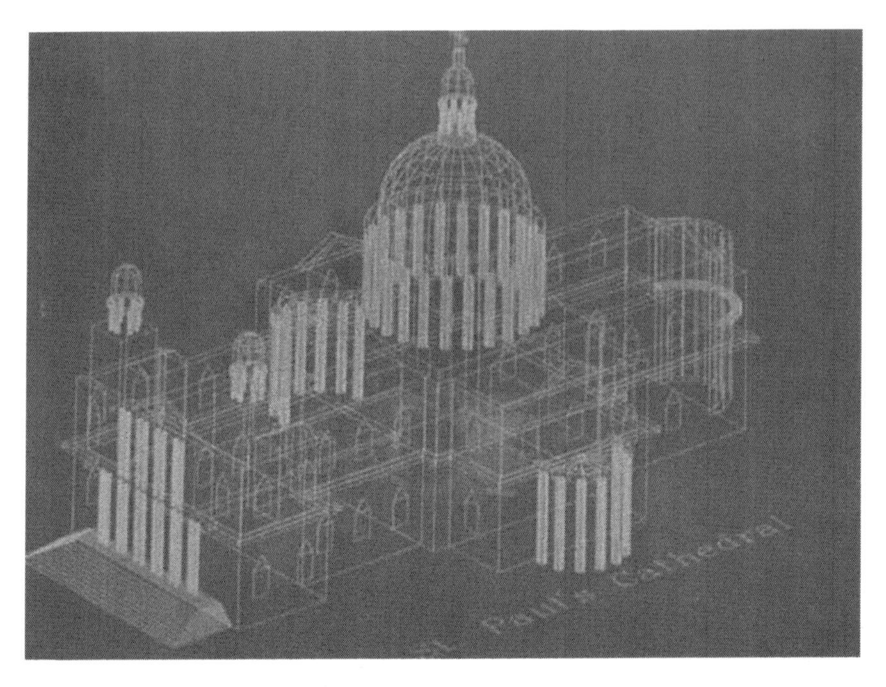

FIGURE 7.13
355 nm laser-etched gold on polymer.

FIGURE 7.14
1 mm thick ferrite plugs laser etched with 355 nm laser.

7.2.2 Atomizers and Nebulizers—Drug Delivery

Atomizers are used to break up a mist into its atomic components for analysis. Nebulizers are used to administer medication in the form of a mist into the lungs. Inhaled aerosol droplets can only penetrate into the narrow branches of the lower airways if they have a very small diameter (<5 μm). Otherwise, they do not travel far into the body and the effects are compromised. Nebulizers use liquid or powdered drugs that are loaded into the device upon usage. Screens with very small laser-drilled holes then break up the liquid and mix it with air or oxygen to make the mist.

There are usually a large number of these holes and the exact size depends on the drug being used and the symptoms being treated. Since they are disposable, the devices must be sold inexpensively. A typical device will have 500–1,000 holes < 5 μm diameter (taper is actually helpful in this case), so fast beam steering is crucial. It has been demonstrated that it is possible to drill over 150 devices per minute (each containing 500 holes) on a roll-to-roll machine using beam splitting and fast beam steering with galvanometers and AODs (acousto-optic deflectors).

7.2.3 Microfluidics

Microfluidics is a multidisciplinary field encompassing physics, chemistry, engineering, flow dynamics, and frequently biology. It deals with the manipulation and flow of gas and liquids in a miniature, typically submillimeter system. Although the flow properties are the same on both a macro- and microscale, factors such as surface tension, viscosity, and even electrical charges become much more important on the microscale because of the much larger surface-to-volume ratio. Even flow inertia, which is much more important on the macroscale and has negligible effect on the microscale, is affected. The channel diameter scale is in the range of hundreds of nanometers to a couple hundred microns. Because of this, larger designs cannot simply be scaled down; instead, new manufacturing techniques are called for in order to exploit the microscale physical properties.

Microfluidics experiments have been ongoing since the 1980s; in the early 1990s, the first major application, ink-jet print heads, was commercialized. The field is generally divided into continuous-flow and digital (droplet-based) schemes; the former is the easiest to implement but the latter offers more flexibility. Common materials include polymers (Nylon™, PET, PMMA, and silicone), sapphire, glasses (quartz, fused silica, Pyrex™, BK7), and Si. The appeal in the medical world, particularly for diagnostics, is that the microchips require only a tiny amount of both sample and reagent for each process, typically nanoliters rather than the milliliters used by conventional plate assays. Also, microscale reactions occur on a much faster time scale and are easily chip automated, requiring little human intervention and thereby reducing costs and potential contamination. Desirable characteristics include sharply defined features, smooth walls, and optically clear surfaces; the parts must be produced with high reproducibility and high speed in order to make the process economically feasible. The features of interest include holes, channels, sample chambers, and cones in both two and three dimensions.

Far-infrared lasers, primarily CO_2, have been shown to be very flexible and fast in the production of polymeric materials. On a dollar-per-watt basis, this laser is much less expensive than other options discussed later, but it does act by thermal vaporization, so the feature quality and attainable feature sizes may be an issue. The best results are obtained in PMMA, which has high far-IR absorption. Features can be produced at relatively high speeds with attainable channel depths between 100 and 300 μm and widths from about 100 to 250 μm. Because of the melt, surface roughness is relatively small (in the micron range), but recondensed polymers on the structure edges can be a problem. Polycarbonate is also a good candidate for CO_2 laser processing.

Visible lasers are only used in specialty applications because they are not short enough in wavelength to produce very fine features or give very clean cuts and not long enough in wavelength to provide cost efficiency and speed. Therefore, most of this work is done with UV lasers. The choices are 355 and 266 nm DPSS (diode pumped solid state) lasers and 248, 193, and 157 nm excimer lasers. These are listed in order of increasing cost and complexity, but decreasing attainable spot size on target. The quality of cut also improves as the wavelength decreases.

Since DPSS lasers have a Gaussian beam and are typically used with galvanometers, programming complex patterns is very straightforward. In fact, by focusing into the material of partially absorbing substrates, fluidic channels within the bulk can be manufactured. Many materials show very low absorption above 200 nm wavelength, so vacuum ultraviolet (VUV) wavelengths of 193 and 157 nm are an option. The former requires at least a purged beam delivery because air—specifically oxygen—absorbs about 50% of the available photons over about a 1 m path length. This wavelength is good for materials such as Nylon™ and PET. The shorter wavelength 157 nm laser can be used for even more difficult materials like polytetrafluoroethylene, quartz, and fused silica, but this requires the use of an evacuated or very well-purged beam delivery system as even a small amount of oxygen present in the beam path will kill the process.

Until quite recently these options were the only ones to consider, but over the past few years commercially available picosecond and femtosecond lasers have proven to be an interesting alternative. The short pulse length introduces nonlinear effects and makes absorption in an otherwise nonabsorbing material possible. These lasers can also be focused into the bulk material to create fluid paths within a substrate.

Figure 7.15 shows microfluidic channels less than 50 μm wide etched into quartz using a femtosecond laser. NASA Ames has recently announced a joint

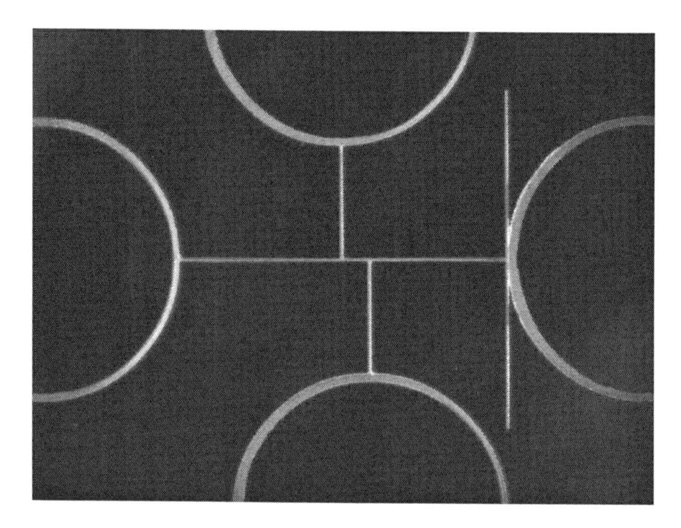

FIGURE 7.15
Microfludic channels in quartz. (Courtesy of Raydiance.)

project for using an 800 fs laser to manufacture next-generation microfluidic devices that will be deployed on free-flying nanosatellites, the International Space Station, and future lunar and planetary research laboratories. Results from this work will advance capabilities for molecular and cellular diagnostics, enable rapid drug discovery and screening, and expand the understanding of the effects of the space environment on biological systems.

Other application areas include acoustic droplet injection, evolutionary/cellular/molecular biology and biophysics, fuel cells, drug delivery, microfilters, "lab on a chip," MEMS, nozzles, biosensors, and gene sequencing. Market areas include defense/home security, medical diagnostics, air-quality monitoring, ink-jets, and other liquid mixing and dispersing. Scientists at Sandia National Laboratories are developing ChemLab™, a portable, handheld chemical analysis system for homeland security, defense, environmental, and medical applications. This unit can detect chemical warfare agents and proteins, as well as biotoxins such as ricin, staphylococcal enterotoxin B, and botulinum. It can also identify viruses and bacteria using protein fingerprinting.

Other companies are working on integrating a host of analytical procedures onto one substrate, the lab on a chip. Biochips have been in the marketplace for many years in various forms, but improvements in the technology are having a revolutionary impact on the next generation of assays. This is especially true of three-dimensional structures assembled from multiple bound layers. A large number of experiments, such as DNA analysis, drug detection, screening of patients, and cancer and other illness detection, can thereby be performed in a single operation and with a very small amount of fluid.

Microfluidics is revolutionizing the way in which many things are done. Processes will become more efficient and chemical reactions that are not currently possible because the reagents are either too unstable or too costly will be routine. Lasers are poised to play a great role in the development of this technology.

7.2.4 Angioplasty and Stents

Angioplasty is the technique of mechanically widening a narrowed or obstructed blood vessel. A collapsed balloon on a guide wire is inserted into the blocked area and then inflated to its uncollapsed size using water pressures of up to 500 times normal blood pressure. This crushes the fatty deposits and thus opens up the blood vessels for improved flow. The balloon is then deflated and withdrawn.

In some cases, very small (about 1 μm diameter) holes are drilled into the balloon so that, after inflation, a restenosis (defined as the formation of new blockages after an angioplasty operation has been performed) inhibiting drug can be delivered locally to the affected area. In other cases, a second catheter is inserted and the drugs are delivered through laser-drilled holes, which are somewhat larger since they do not have to stand up to the high pressures of the inflated balloon. Peripheral angioplasty refers to the use of a balloon to open blood vessels outside the coronary arteries, and this procedure is usually used in conjunction with stents—artificial tubes set into the body to counteract localized flow restrictions. The size of these stents depends on where in the body they are to be placed, but for coronary operations they are usually less than a few inches long. Since the stents are inserted through the femoral artery, they must be flexible enough to travel through the body without causing damage along the way, yet they must be rigid enough to hold the orifice open.

Metal stents have been used for years and there are hundreds of near-IR lasers making metal stents. The problem is that if restenosis occurs, a surgical procedure is required to remove the stent and correct the problem. Unfortunately, this occurs in a large number of cases. Also, postprocessing is needed in order to finish and polish the final product.

One of the ways around this is to use bioabsorbable stents. These devices are manufactured so that they stay intact in the body for a few weeks and then dissolve; if a follow-on procedure is needed, it can be done by a simple balloon expansion. Bioabsorbable stents, however, are extremely heat sensitive, so even UV lasers do not do a good job; ultrashort pulse lasers must be used to get the best edge quality and required cleanliness. Figure 7.16(a) shows a gold metal stent and Figure 7.16(b) shows a bioabsorbable stent laser processed using a femtosecond laser. In both cases, a fixed-beam delivery system is used with coaxial gas assist and a high-precision rotary stage with dynamic motion.

(a)

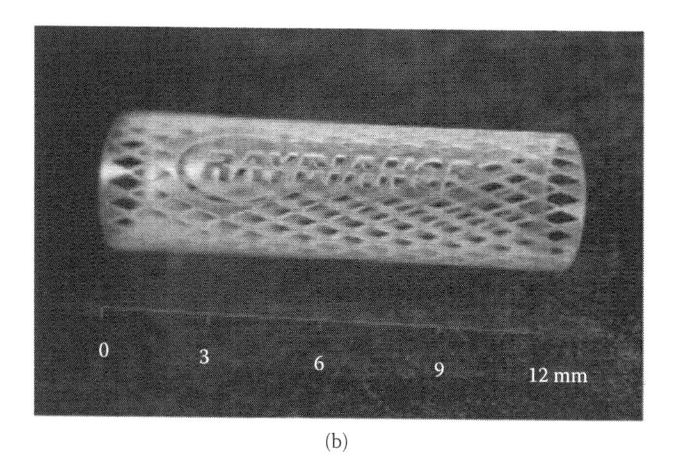

(b)

FIGURE 7.16
(a) Femtosecond laser-processed gold metal and (b) PTGA bioabsorbable stents. (Courtesy of Raydiance.)

7.2.5 Catheters—Drug Delivery

A catheter is a flexible tube that is inserted into the body to allow for drainage, localized drug delivery, or access by surgical instruments. Catheters are sometimes laser marked with graduations so that the surgeon can determine precisely how far the insertion has progressed at any time. Drainage catheters do not require the small hole sizes and tight tolerances that drug delivery catheters require, and the holes can usually be made by mechanical methods. However, drug delivery catheters require precise hole size and positioning so that the minimal amount of drug can be delivered locally without damage to the surrounding area.

A range of polymers can be used as long as the material has biocompatibility, but the full device usually includes some molded plastic components and, sometimes, a guide wire or needle. Drug delivery catheters are generally meant to be used and then removed from the body. Holes can be placed longitudinally, spirally along the length, or in a number of other different patterns; different diameter holes can be found on different portions of the device. Sometimes other areas of the catheter are also laser machined (e.g., the tip). UV and/or short pulse lasers are used in this application.

7.2.6 Microfilters

The best example of a microfilter for medical applications is a blood filter. Silicone-based filters are constructed by drilling a number of 3.5 μm diameter holes, which selectively trap the white blood cells while allowing the red blood cells to pass freely. All of these microfilter applications have a few things in common. First, the size of the holes is in the range of a few microns to even submicrons. Filters generally use through-holes, but blind vias can also be drilled with submicron depth control and hole diameter repeatability. Also, aspect ratios of better than 10:1 can easily be achieved, and the range of materials includes polymers, glass, ceramics, and metals. Figure 7.17 shows an array of small holes (10 μm) drilled using an excimer laser and multiple hole imaging mask. By stepping and repeating, this procedure can produce very large filters on the order of 300 mm across.

7.2.7 Transdermal (Patch/Perforations)

Transdermal patches are polymeric formulations that, when applied to skin, deliver a drug at a predetermined rate to achieve systemic effects. Transdermal dosage forms are becoming popular because of their unique advantages, such as controlled absorption, uniformity, improved bioavailability, reduced side effects, painless and simple application, and the ability to terminate drug delivery simply by removing the patch from the skin.

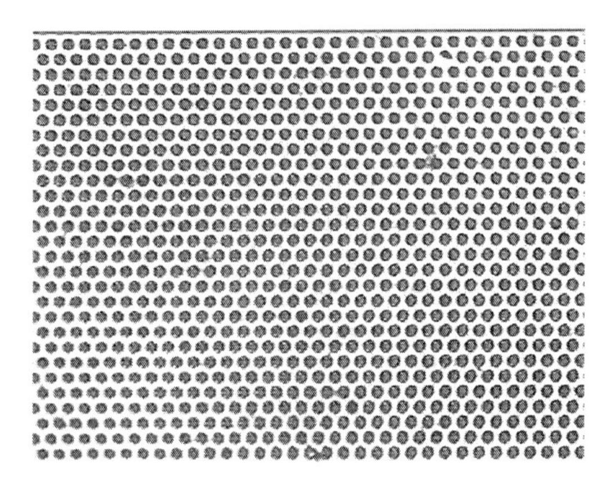

FIGURE 7.17
Excimer laser-drilled biofilter.

Unlike 90% of medication, these patches allow controlled release of a drug over time. Drugs that are released rapidly (taken orally or injected) produce a quick and high concentration in the body, followed by a sharp decline. At a level that is too low, the drug is not effective; at a level that is too high, there may be undesired side effects. The objective is to maintain a steady dosage over a long period of time.

Lasers are used to drill the microholes in the polymers that allow the drugs access to the skin when a release liner is removed and the patch is applied. Some examples are for use by smokers (nicotine), for seasickness, and for estrogen delivery. Another somewhat related application is the use of lasers to drill small holes in pills with a water-permeable jacket that has been laser drilled with one or more holes so that, as the tablet passes through the body, the osmotic pressure of water entering the tablet pushes the active drug through the orifice.

7.2.8 Fluid Metering Devices—Orifices

A good example of a fluid metering device is a plastic injection molded part with a laser-drilled hole for drug delivery. Injection molding is a rapid and relatively inexpensive way to make high volumes of disposable medical products, but the limitations on the molding process preclude molding holes small enough to serve as drug-metering orifices, so lasers are used to drill these holes. First, the proper laser wavelength must be chosen so that a clean hole is drilled. Second, the required hole size must be achievable. UV lasers are the best candidates for most of these types of applications.

Figure 7.18 shows a UV-drilled 25 μm hole in a polycarbonate injection molded part from the top (a) and side (b). Drilling rates of 1,000 parts in 40 minutes were achieved using an excimer laser with a split beam, eight-up delivery system (drilling eight parts simultaneously). However, this process required

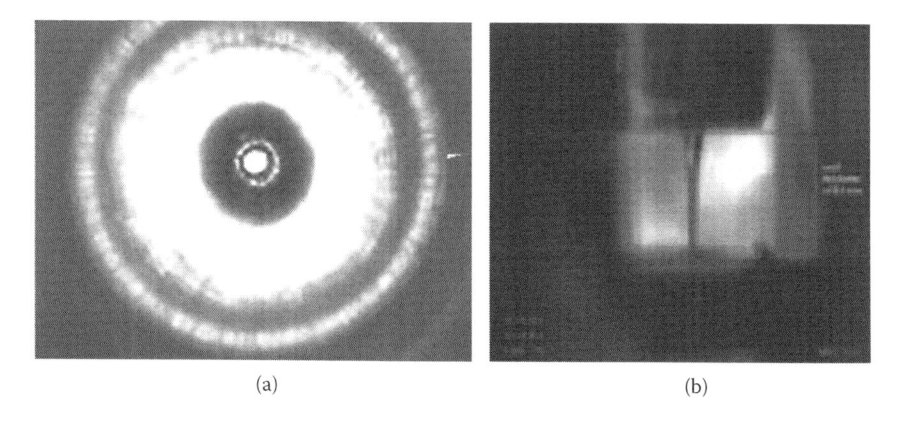

(a) (b)

FIGURE 7.18
UV laser-drilled hole in plastic injection molded part: (a) top view; (b) side view.

three people to run the laser (load, unload, and inspect) and the large number of optics required a high skill level of the operators. This process was later transferred to a 266 nm laser with a single beam delivery. This laser was used because longer wavelength lasers (355 nm) did not produce the desired results.

Figure 7.19 shows a 266 nm laser tool designed for the production floor that uses a bowl feeder and tooling to orient and front surface locate the parts at a precise point in space so that a single part can be drilled. While this process takes over 1 hour to drill the same 1,000 parts, it runs unattended. Also, using a rotary indexer, it is possible to inspect every part after laser drilling instead of taking random samples. Testing is done using a calibrated flow meter and air. This reading must be correlated to liquid flow. (Liquid is not used in the testing, so the parts remain clean and dry.) Since gas flow is much more dependent on temperature, humidity, elevation, and other things, the testers are calibrated to a set of "gold" standards that have been measured with liquid and give acceptable results.

7.2.9 Cutting Flat Sheet Stock

In a sense, this is one of the easiest applications. A simple CAD program directs the laser to flat stock stretched over a vacuum chuck and parts are

FIGURE 7.19
266 nm laser on production floor.

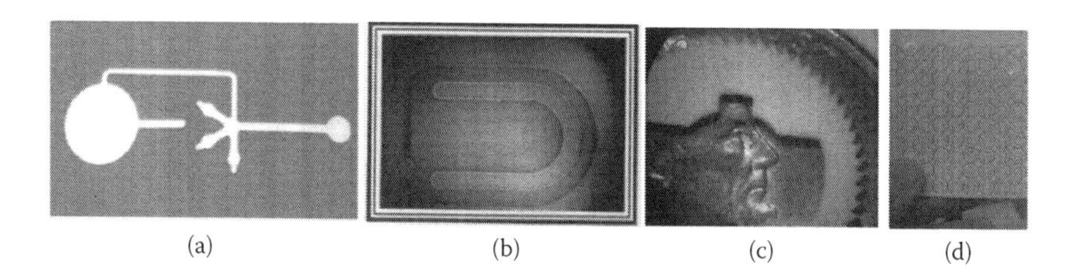

FIGURE 7.20
Different products cut from flat sheet stock. (a) cutout of laminate for hemocytometers; (b) 3 mil thick laser patterned Mylar™; (c) small gear compared to a penny; (d) thickness gauge washers made from shim stock.

cut using a step-and-repeat process. After laser processing, the parts are separated from the chad or, if tabs are used, the whole sheet is delivered. Figure 7.20(b–d) shows photos of three such products. The cutout shown in Figure 7.20(a) is from a hemocytometer made from a laminate. The part is about 1 in. long. Note the clean edges. The cutout shown in Figure 7.20(b) is made from 3 mil thick Mylar™. It is about 13 mm across and is used as an air flow regulator. The cutout shown in Figure 7.20(c) is a small gear in rigid plastic made for an implantable mechanical device. The cutout shown in Figure 7.20(d) is simple shim stock bought at a hardware store and laser cut to make thickness gauge washers. In this case, small tabs are left on the parts so that handling is easier.

7.2.10 Three-Dimensional Surface Structuring

Since UV and ultrashort pulse lasers only remove a fraction of a micron per pulse, it is possible to use them to etch three-dimensional patterns with very high resolution. In order to make a real three-dimensional structure, sophisticated software is needed to take a three-dimensional file and convert it into machine language to control the placement and dosage of the laser on the part. It is also possible to do structuring by using simple masking or feature generation.

Figure 7.21 shows a structure that was generated using a 248 nm excimer laser and a mask with "diamonds" allowing the light to pass through the streets between the diamonds and etch the part. The etched lanes are about 10 μm wide. This device was used as a small chromatography column for pregnancy testers. IR lasers can also be used for surface structuring, usually on metals.

7.2.11 Marking

As stated previously, in many cases, marking is not really material removal, but it is a large market for lasers, especially in the medical field. Traceability is key in medical device manufacturing, so marks (alphanumeric, bar code,

FIGURE 7.21
Three-dimensional structure in polyimide.

and two-dimensional matrix code) are used extensively just for identification purposes. Figure 7.22 shows two bar code marked products. The first is an identifier on a catheter. The second is an identifier on a polypropylene sperm storage vial; coding directly on the container can minimize the chance of mixed up paperwork and surprising results!

Marks can also be used as a part of the functioning product. For instance, it is possible to put graduations all the way around the circumference by using a "hot dog roller" configuration and spinning the part while the laser dwells for some time at each graduation location. Then, the part is stopped and alphanumerics are placed on one side to identify the graduation lines. In this case, the marks are part of the functionality of the device and not just an identifier. Also, some sophisticated marking applications utilize ultrashort

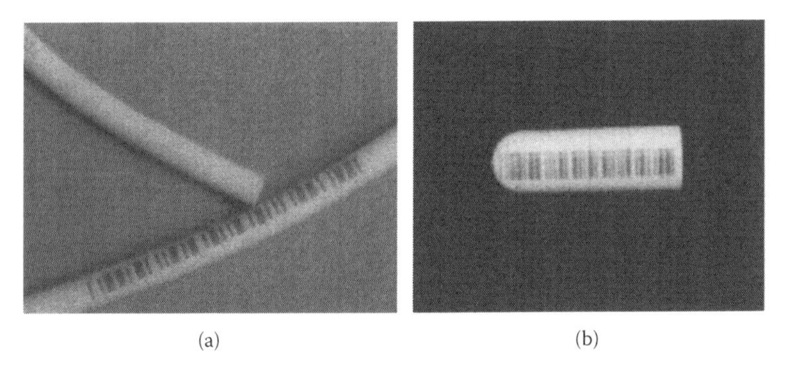

(a) (b)

FIGURE 7.22
(a) Bar code on catheter; (b) bar code on polypropylene vial. (Both photos courtesy of Coherent.)

pulse lasers to mark underneath a thin, transparent outer layer. To date, this is perhaps the largest nonmedical industrial application of picosecond lasers. Hundreds of such systems are deployed across factories in mainland China.

7.3 Defense/Aerospace

A host of applications using lasers in the defense and aerospace fields cannot even be discussed because of their proprietary or sensitive nature. While outside the main scope of this book, lasers are nevertheless used for welding, drilling, and cutting thick metals and composites; marking; and repair. The big application is in the manufacture of jet aircraft engine parts, but other hot topics include sensors, RFIDs, range finding, missile guidance, and defensive lasers.

7.3.1 Cutting and Drilling Composites—Carbon or Glass Fiber

The most common example of a nonhomogeneous material is FR4, which is an epoxy resin embedded with glass fibers. This material is used in printed circuit boards and a host of other electronic and mechanical devices. It has many material advantages, such as relatively low cost, mechanical rigidity, excellent dielectric properties, and water resistance. It can be drilled with far-IR and UV lasers, but the difficulty is in dealing with the nonhomogeneity.

Glass fibers ablate at a much higher energy density than the epoxy and this higher required energy can sometimes cause problems with the epoxy in the form of undercutting or burning. The real difficulty is that the fibers are not distributed uniformly. For instance, if a large number of holes are required to be drilled, there will be areas where there is no glass, areas where there may be a single glass fiber, and areas where there are crossed bundles of fibers. All of these cases must be accounted for and, in a worst-case scenario, the drilling speed for the panel will be slower because all the holes have to be drilled using the parameters for drilling through fiber bundles rather than pure epoxy.

Carbon fiber epoxy material is used in engine components for noise reduction spoilers. This material has some of the same problems as FR4, but the carbon fibers also tend to propagate any heat generated by the laser process, so care must be taken to avoid burning of the epoxy not only around the holes, but also in areas removed from the actual laser processed area. Ultrashort pulse lasers are finding applications in the processing of composites because they can ablate almost any material and have minimal thermal side effects. This provides a distinct and unique advantage over other methods. Figure 7.23 shows a laser-drilled carbon fiber epoxy panel compared to a

FIGURE 7.23
Laser-drilled carbon fiber epoxy.

pencil. This material can be up to several millimeters in thickness, although most work is done on panels of less than 1.5 mm.

7.3.2 Wire Stripping and Marking

Wire stripping has been discussed in Section 7.1.6; the same comments apply to wires used in aerospace applications. Marking of aircraft wire is also done using lasers. Teflon™ insulating jackets are used extensively in aircraft and this material is very hard to mark using ink-based technologies because the ink does not stick. While UV lasers sometimes have a hard time machining this material, it marks extremely well. Figure 7.24 shows a UV laser-marked aircraft wire.

7.3.3 Hole Drilling in Aircraft Engine Components

Modern jet engines have up to hundreds of thousands of holes drilled into various components such as turbine blades, nozzle guide vanes, combustion chambers, and afterburners. These holes are less than 1 mm in diameter; some are through-holes and some are shaped holes. Near-IR lasers (high pulse energy Nd:YAG or fiber lasers) are typically used in either a percussion or trepanning mode. These holes allow a film of cooling air to blanket the components; this extends life, reduces maintenance, and achieves superior performance characteristics.

Drilling is done on a variety of metals and exotic alloys; important considerations are drill time (remember that there is a large number of holes), recast layer, oxidized layer, taper, and microcracking. Electrode discharge

FIGURE 7.24
UV laser-marked aircraft wire.

machining (EDM) can be used for the metals, but there are some significant drawbacks, such as the difficulty in making angled holes, the need to use lots of liquid coolants, and consumption of the electrodes. The biggest drawback, however, is that many components are coated with a heat-insulating layer of zirconia ceramic that acts as a thermal barrier coating (TBC); this coating is nonconductive, so EDM will not work on these components.

Hole quality is judged based on entrance and exit hole diameters, taper, roundness, oxidation, and recast layer. If straight-through holes are needed, percussion drilling with a fixed-beam delivery system is the fastest way to accomplish this. However, shaped holes are also used and these shapes can most readily be made using galvanometer-based beam delivery. Usually, the making of shaped holes is a two-step process where the initial hole is drilled using percussion drilling and the entrance is shaped using galvos. These components are large and have complex shapes and the process requires multiple axis stages and/or robots.

Figure 7.25 shows a typical laser-drilled turbine blade. Figure 7.26(a) shows an array of shaped holes on a flat substrate; Figure 7.26(b) shows a laser-drilled vane. Figure 7.27 shows the near-IR laser in the process of metal drilling.

7.3.4 Removal of Thermal Barrier Coatings

Turbine airfoils are very expensive to manufacture; they become damaged in use quite frequently, so refurbishment is common. Currently, the refurbishment process requires the use of noxious chemicals and involves a multistep process. It has been shown that lasers can be used to remove the outer layers down to the superalloy cleanly and cost effectively for subsequent recoating.

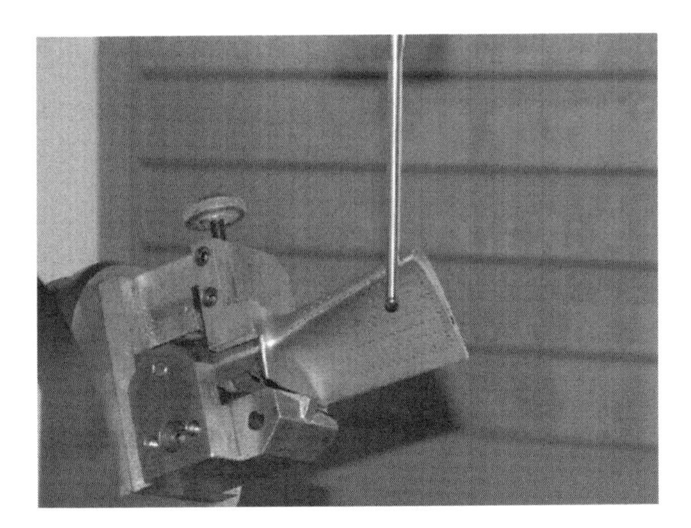

FIGURE 7.25
Laser-drilled jet engine turbine blade. (Courtesy of CCAT.)

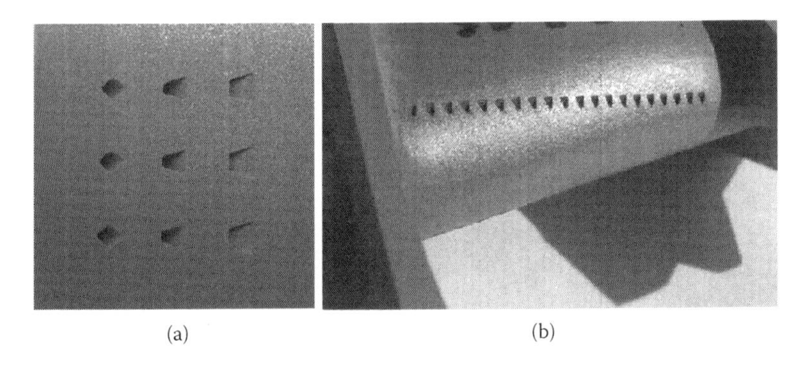

(a) (b)

FIGURE 7.26
Shaped holes: (a) on flat stock; (b) on actual engine vane. (Courtesy of CCAT.)

On a typical turbine airfoil is a layer of electron beam physical vapor-deposited TBC. The TBC is yttrium-stabilized zirconia, commonly referred to as 7YSZ. There are four layers of material on the turbine airfoils: the outer TBC layer, a layer of aluminum oxide (α-Al_2O_3), a layer of platinum/aluminum (Pt/Al), and, finally, a superalloy. All the turbine airfoils are taken from an engine after many hours of use; however, some of the turbine airfoils are taken from the engine without cleaning, and other turbine airfoils are cleaned to remove engine deposits. The two types of airfoils are easily distinguished, as illustrated in Figure 7.28(a) and (b) showing an as-removed turbine airfoil and a cleaned airfoil, respectively. The dark brown material in (a) is debris; most (but not all) of the debris is removed by the cleaning process. It should also be noted in (b) that a significant amount of underlying

FIGURE 7.27
Laser drilling of engine components. (Courtesy of CCAT.)

metal can be seen on the cleaned airfoil. Some of the underlying metal is exposed by spallation (lower left); other metal (upper left) is exposed by differential wear or by the cleaning process.

Lasers must be capable of completely removing all of the outer layers without damaging the underlying metal, and they must be able to compete with existing processes regarding processing time and cost. Furthermore, the complex shape requires sophisticated part handling. It turns out that many

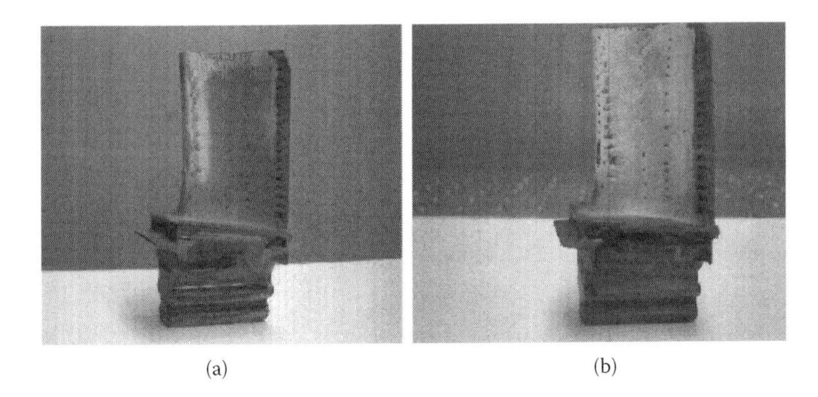

(a) (b)

FIGURE 7.28
(a) Uncleaned engine turbine airfoil; (b) cleaned engine turbine airfoil.

lasers can remove the TBC; however, avoiding damage to the underlying material, removing everything completely, and doing all of this economically are the challenges.

7.3.5 Thin Film Processing (Large Panel Format)

Large panel formats are sometimes needed for esoteric applications. For instance, some proposed space antennas are as large as a football field! It is unrealistic to try to build a motion system to accommodate this size, so smaller sections are made and "tiled" together. Even the "small" formats are on the order of 2 × 10 ft. This size motion system is commercially available and affordable, though not cheap. A panel of this size can be made on a 2 ft stage (which is common) by precise indexing and camera alignment combined with sophisticated software. Figure 7.29 shows a laser system modified to handle large formats. The material is indexed between rolls as shown.

7.4 Renewable Energy

This is potentially the most lucrative application area for lasers in the near future. In the renewable energy (or even energy conservation) areas, lasers are used from the far IR to the UV in many different aspects of manufacturing. In the case of energy production, lasers are used to manufacture the devices that provide the power (for instance, solar panels) and that store the power (batteries). The third leg of large-scale renewable energy production is transportation, in which lasers play no immediate role—but who knows about the future?

FIGURE 7.29
Modified laser system to process 2 × 10 ft panels.

7.4.1 Light-Emitting Diodes

There are currently approximately 12 billion incandescent lightbulbs in use globally; they account for about 40,000 trillion hours and 1 billion tons of coal annually. In the United States, the energy required to service lighting needs accounts for about half of the energy used by all automobiles. Light-emitting diode (LED) lighting can save around 80% over incandescent lighting. LEDs have high efficiency (>150 lumens [lm]/W) and long, useful lifetimes (>50,000 hours) and they do not contain hazardous materials like mercury. The efficiency of a typical incandescent bulb is about 13 lm/W. The current cost for incandescent-based lighting, however, is about $1 per 1000 lm, while LED-based lighting costs about $10 per 1000 lm.

The cost of LEDs, however, is dropping rapidly as new manufacturing techniques—including laser processing—are brought online. Within a few years, through cost reduction and government regulation/subsidy, LEDs will have a cost parity with incandescent bulbs, which opens up a $100 billion market. LEDs are currently used in mobile handsets, projectors, TVs, flashlights, headlights, and a host of other devices, and the consumption of this product is rapidly growing.

The main application areas for lasers are in scribing, "liftoff" and marking. UV lasers are used to make clean, fast, and narrow scribes with front side processing that produces up to 35% higher yields than mechanical processes. Cycle times are also reduced from hours to minutes. Typically, either a 355 or 266 nm DPSS laser is used and scribe speeds are 100–300 mm per second, depending on scribe depth. The 532 nm wavelength can also be used on some substrates. The laser sources require high energy per pulse, good beam stability, high pulse repetition rate, good beam quality, stable laser performance, and high up-time. This process works on sapphire, SiC, GaAs, InP, and GaP.

Figure 7.30 shows a 2.5 μm wide laser scribe and Figure 7.31 shows a scribed GaAs wafer. In recent years, dicing using ultrashort pulse lasers has been adopted in many factories. Here, the ultrashort pulse beam is focused inside the wafer, creating an internal fracture. This fracture can be used to break the wafer mechanically just as in the case of scribing. Internal fractures have the advantage of not leaving any traces behind and there is no debris such as one sees with scribing.

LED liftoff is done on a large area with an excimer laser. It can be fast with relatively low operating costs and low-temperature operation. It not only reduces waste, particularly sapphire, but also results in increased LED output. The basic concept involves imaging a 248 nm beam through a transparent substrate (sapphire) and "lifting" gallium nitride off the back side. The running time is better than 1 minute for a 50 mm diameter wafer. The same principle applies to other substrates and coatings, perhaps using different wavelengths (193 nm for aluminum nitride). The process requires a very homogeneous beam, and wafer preparation is important in order to avoid fractures. Figure 7.32 shows a schematic of LED liftoff and an actual substrate where nine dies have been processed with one pulse.

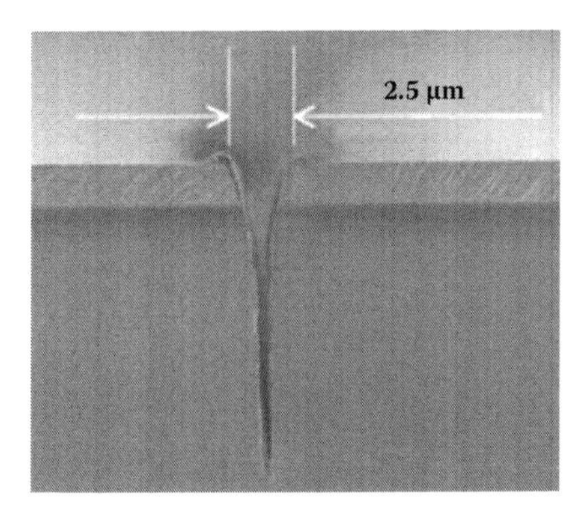

FIGURE 7.30
2.5 μm wide laser scribe. (Courtesy of JPSA.)

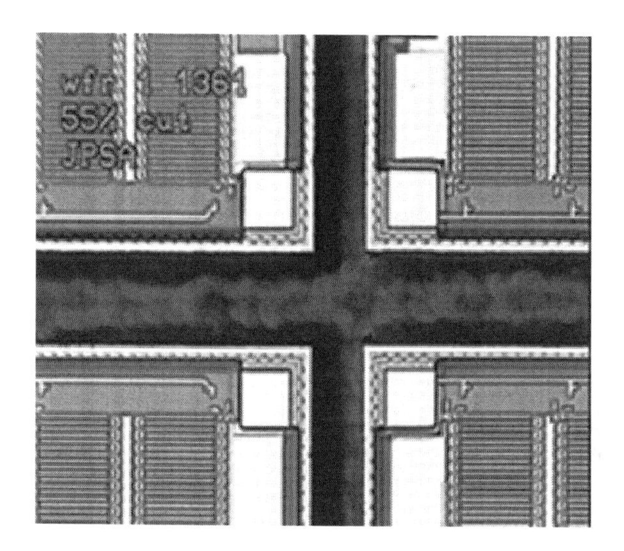

FIGURE 7.31
Scribed GaAs wafer. (Courtesy of JPSA.)

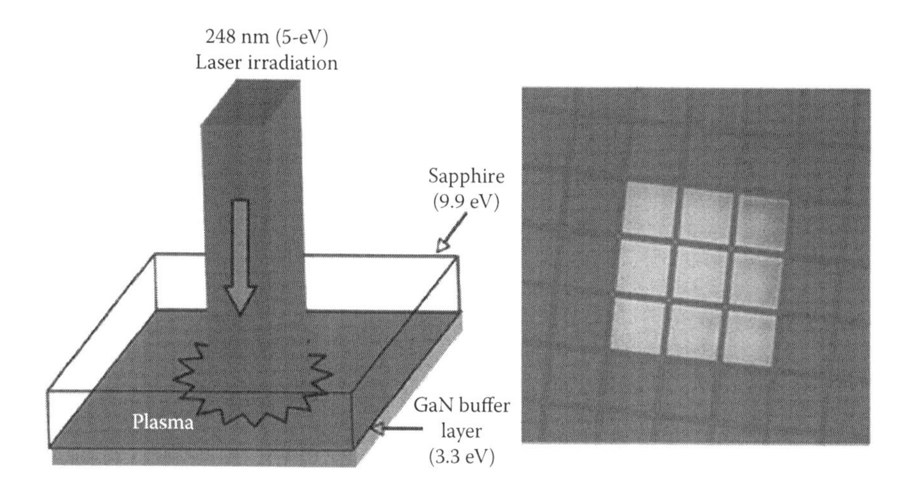

FIGURE 7.32
Schematic of LED liftoff and the resulting nine-die liftoff with one laser pulse. (Courtesy of JPSA.)

7.4.2 Batteries

The market for lithium batteries is expected to top $40 billion in 2012 and has grown by 90% since 2007. There is currently a target of 1 million electric cars on the road in the United States in 2012. Lasers, primarily near-IR lasers, are used in the joining of dissimilar materials, sealing battery housings, and joining internal components to electrodes and terminals. They are also used in cutting the electrodes and other small components. New generations of batteries have numerous thin metal plates in each device and these plates are sometimes delivered on a roll to enhance manufacturability. Using lasers is an ideal way to pattern large numbers of plates rapidly on a continuous feed roll.

7.4.3 Microtexturing for Friction Reduction

While not strictly renewable energy, microtexturing is similar to LEDs in that it is an energy-saving application. Reducing the contact surface area on moving parts decreases the energy consumption and also makes for longer lifetime of components. The basic concept of microtexturing is to pit the surface in a carefully controlled manner. This is done by exploiting laser technology in such a way that surface flatness is not affected and there is no danger of creating stress raisers or cracks. The laser removes material from the surface, creating tiny dots or grooves, according to the specific requirement. For example, a dot might be 10–200 μm in diameter with a depth of up to 100 μm, but this surface manipulation has no specific geometry; it is always application specific. The advantage that can be gained through surface microtexturing is not due to enhanced oil retention, in the way that a piston ring working face might be micropitted to retain oil, but rather it causes a change in the way the oil flows between two surfaces.

7.4.4 Fuel Cells

This is another application area, like batteries, that is expected to undergo explosive growth in the next few years. Also, like battery production, near-IR lasers will particularly play a very big part on the joining and additive manufacturing side. Specifically, lasers are used to join metallic interconnects, which are a fundamental element in fuel cells, and millions upon millions will be required as fuel cell technology is widely accepted. A single fuel cell can have hundreds of feet of welds, so the process must be fast and inexpensive, which favors IR lasers.

7.4.5 Thin Film PV (Photovoltaic)

In 1 hour, enough sunlight hits the earth to power the entire annual energy need of the planet! If only a fraction of the energy hitting the earth can be harnessed, energy will be freely available without all of the negative side effects of carbon-based energy generation. Crystalline solar cells still dominate the market because they are higher in efficiency, but thin film cells are cheaper to produce and they can more easily be scaled to high volume. Although lasers are not used extensively in the manufacture of crystalline solar cells, they are used extensively in thin film solar applications because of lower fabrication costs, savings in Si consumption, and superior performance resulting in lower cost per unit output. Lasers surpass mechanical scribing in quality, speed, and reliability. The cost is still higher than "grid" costs, so the Holy Grail for thin film solar applications is grid parity. The biggest use of lasers is in making the P1, P2, and P3 scribes.

Thin film devices consist of multiple layers of films deposited on glass. The following steps are applicable to all thin film production including a-Si (amorphous Si, the most common), CdTe (cadmium telluride), and copper indium gallium selenide (CIGS), although CIGS has some differences that are discussed later. The total stack-up consists of (from the incoming sunlight side down) a transparent protective layer (glass), a transparent conductive layer, an active layer, a metallic conductive layer, and, finally, another glass protective layer, as in Figure 7.33.

First, the glass is coated with a TCO a few hundred nanometers thick that forms the front electrodes. This layer is then scribed (P1 scribe) through the glass using a 1064 or 1070 nm laser, as shown in Figure 7.34. Processing

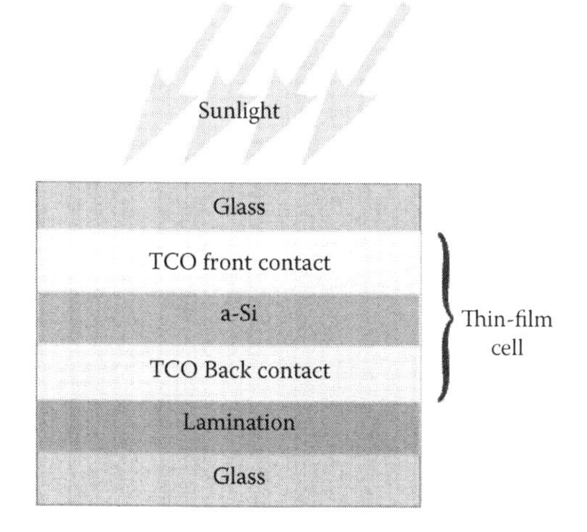

FIGURE 7.33
Typical thin film solar cell stack-up.

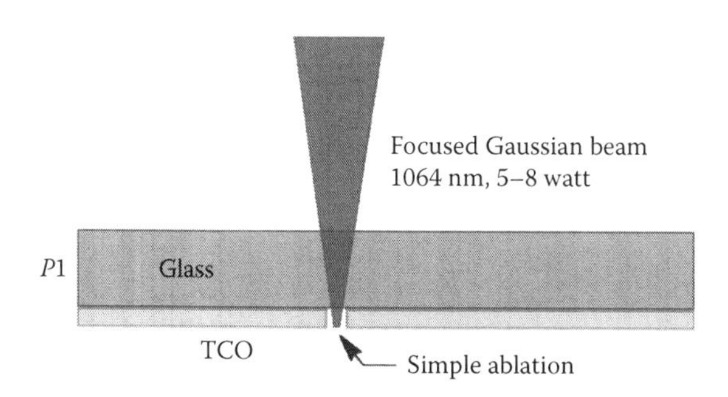

FIGURE 7.34
P1 scribe.

through the glass has a few advantages: (1) Debris will fall down and away from the substrate, and (2) the energy-per-pulse requirement is much lower since only the interface, and not the entire layer, is ablated.

Next, Si is deposited to 2–3 μm thickness and is then scribed (P2) using a 532 nm laser (so that the TCO is not touched), as shown in Figure 7.35. Finally, a thin metal layer (rear electrode) is deposited to submicron thickness and then scribed, again using a 532 nm laser, which removes both the Si and the metal electrode, as shown in Figure 7.36. High volume and low unit cost are required. The scribes must be high quality with no defects. The area between P1 and P2 is not active, so it must be minimized. Scribes of tens of microns width and hundreds of microns offset are common, but better efficiency is obtained by minimizing both.

Narrow scribes (~25 μm) are ideal, but this means that the lines must be straight with no wandering and no defects. Excellent cut quality with no edge roughness or peeling is required since defects negatively influence solar conversion. Required scribe rates are several meters per second; no active depth control is possible, so the process must be self-limiting. Laser characteristics include high repetition rate (>100 kHz), short pulse length,

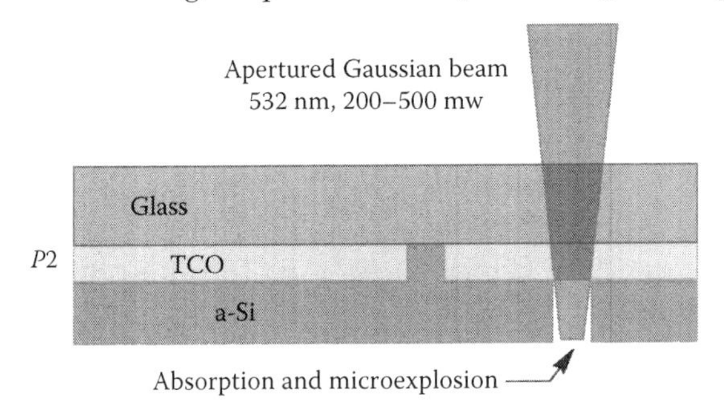

FIGURE 7.35
P2 scribe.

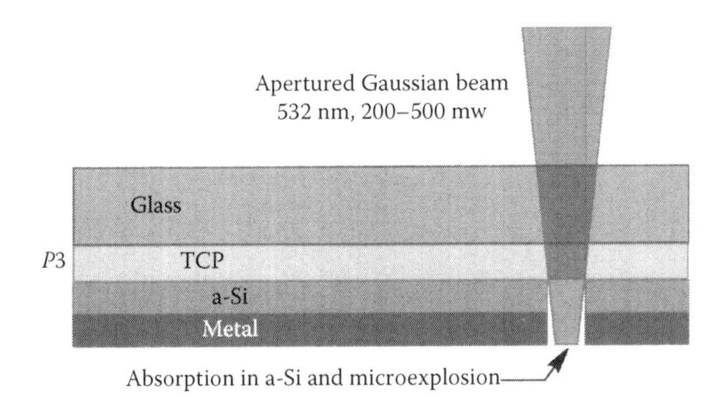

FIGURE 7.36
P3 scribe.

excellent pulse-to-pulse energy and beam pointing stability (<50 μrad), and high reliability/low maintenance. Picosecond lasers are also becoming more attractive as the layers become thinner and collateral thermal damage to the neighboring material must be avoided.

7.4.6 Copper Indium Gallium Selenide

In CIGS manufacturing, lasers are typically only used to remove the P1 layer of Mo. CIGS has generally proven difficult to scribe with nanosecond lasers since the layers are over 1 μm in thickness and thermal damage results. For this reason, mechanical processes are currently widely used. Sophisticated ultrafast lasers with scribe speeds of up to 15 m/s have been demonstrated as a potential solution, but these lasers are not considered to be either cost effective or robust enough for 24/7 manufacturing. However, in spite of manufacturing difficulties, CIGS does have potential for high conversion efficiency: 20%, which competes with crystalline solar cells. For this application in particular, picosecond lasers are being deployed in large quantities and it is expected that large-scale manufacturing of CIGS using picosecond lasers will be the industry norm.

7.4.7 Edge Deletion

Edge deletion removes all the coatings within about 10 mm from around the edges of a thin film solar panel. This is done in order to seal against humidity, which kills a solar cell's efficiency. Pulsed lasers eliminate consumables used, for instance, in sandblasting. A large quantity of material is removed, so the lasers need to accommodate this requirement with clean removal, high depth of field (>1 mm), and high TAC (Total Accumulated Cycle) times. This means that the lasers need high beam quality and high power—making this an ideal application for near-IR lasers. Pulse-to-pulse stability is also crucial since each pulse makes an electrical isolation. High repetition rate is more important in this application than high pulse energy.

7.4.8 Emitter Wrap-Through and Metal Wrap-Through

Emitter wrap-through (EWT) and metal wrap-through (MWT) involve drilling a large number of holes very rapidly through Si in crystalline solar cells to improve efficiency. In modern PV production, a throughput of 1,800–2,000 wafers per hour is common; this introduces strict requirements on the drilling, so lasers are a good choice. For instance, for MWT drilling on a 6 × 6 in. wafer, about 100 holes of diameter 50–100 μm must be drilled within about 2 seconds. Using a laser to drill the holes is favored because it introduces energy selectively at the drilling position without any mechanical contact of the drilling tool. Because there is no tool wear and laser emission can be controlled precisely, the laser drilling process is very reproducible—a demand of factory automation. The ablation rate can be optimized if the laser pulse duration is in the range of hundreds of nanoseconds or microseconds for near-IR lasers.

FIGURE 7.37
Multiple lasers removing conductive ink in a roll-to-roll process.

7.4.9 Organic PV

Conductive printable inks can be printed on plastic substrates to make very inexpensive products. The efficiency is not high compared to other technologies, but the cost is very low. Very inexpensive fiber lasers (<10 W) are used to pattern the conductive ink; in this case, the laser goes through the transparent plastic and removes the material from the back side. This is done in a roll-to-roll process with the lasers making linear streets as the line moves. Each laser makes one street, so numerous lasers are used. Figure 7.37 shows a production line with over 50 lasers working simultaneously (some of the lasers can be seen in this photo). This application is almost like printing money!

7.5 Other

7.5.1 Automobiles

Lasers are used extensively in automobile manufacturing—primarily in sheet metal cutting and welding, which fall outside the strict definition of micromachining. Nevertheless, many of the electronic components are made using laser technology. Lasers are also used in the manufacturing of the sensors used in "smart" vehicles. Another example is in the manufacturing of automotive fuel filters, which require small holes through up to 1 mm of stainless steel; these holes can be drilled using a high pulse energy near-IR laser at the rate of 120 holes/second. Another application is laser drilling fuel injectors for gasoline and diesel. Gasoline injector holes are drilled in relatively thin metal and can be angled. Lasers are particularly useful for short

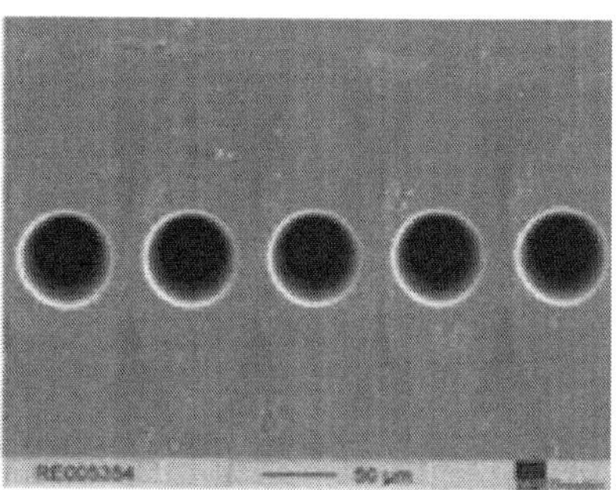

FIGURE 7.38
Laser-drilled ink-jets.

runs and R&D since hard tool costs can be avoided and a part can be made within hours of design. Diesel engines require drilling holes with diameters down to 60 μm in about 1 mm of hardened steel.

7.5.2 Ink-Jets

Ink-jet cartridges have many small holes that are used to deliver the ink. Very high tolerances are needed and excellent hole-to-hole uniformity is required since the human eye is a wonderful detector and can differentiate very slight anomalies. One technique uses an excimer laser to drill all the holes simultaneously using a multihole mask. The setup is expensive and complicated, but large numbers of parts can be made in a very short amount of time. Hole diameters down to 20 μm are common in materials like polyimide and polysulfone. Figure 7.38 shows the nozzles of an ink-jet cartridge with a high-resolution image of the drilled holes; notice the high uniformity.

7.5.3 Cutting and Scoring Display Glass

The traditional method for scoring and breaking glass involves using a mechanical scribe and then a "chopper bar" to break the glass along a line. First, this technique can only produce straight lines. Second, the mechanical contact is problematic when dealing with very thin glass, as is used in current displays. Laser glass cutting addresses these problems by being noncontact and producing no residual stress, resulting in higher edge strength and no chipping or microcracking. Laser scribing is used for glass in the range of 0.3 mm and up, while laser through-cutting is used for glass under 0.3 mm in thickness. In both cases, the laser of choice is CO_2. A special type of glass, called Gorilla™ glass, cannot be cut with the CO_2 laser; forming this glass is a huge challenge in the display glass industry.

8

Materials

While this text is primarily about lasers, this chapter gives descriptions of some of the most common materials that are laser micromachined. It is not intended to be an exhaustive review of any material. Rather, it is simply general comments for the reader who might not be familiar with some of the materials and where or how they are used, how they interact with laser beams, and potential safety considerations.

8.1 Metals

The choice of lasers to process metals depends on which metal, required feature sizes, thickness, speed, and cost. The nominal 1 μm wavelength of Nd:YAG and fiber lasers is ideal for coupling to most metals and these lasers are used in most metal cutting, drilling, and welding applications. However, for thin materials and very small feature sizes, UV lasers are also used. As a general rule, diode pumped solid-state (DPSS) lasers are not very useful for metals over 500 μm in thickness. In the thickness range below 125 μm, they can be fast, efficient, and cost effective, but as the thickness increases, the process gets slower and taper becomes greater. Doubling the thickness more than doubles the processing time. Up to 250 μm thickness, the process slows somewhat but, depending on the application, it can still be cost effective. Up to 500 μm, the process can still be done, but usually requires a higher power laser: 10–20 W. With more than 500 μm thickness using a UV, DPSS laser is achievable only in some cases.

8.1.1 Stainless Steel

Stainless steel is a steel alloy with a minimum of about 11% and up to 26% chromium content by mass; it sometimes contains (depending on the grade) percentages of Ni, Mo, C, and Mn. Stainless steel does not stain, corrode, or rust as easily as ordinary steel, but it is not completely stain-proof. Different grades and surface finishes are available and it is used when both the properties of steel and resistance to corrosion are required. Unprotected carbon steel rusts readily when exposed to moisture.

On the other hand, stainless steels contain sufficient chromium to form a passive film of chromium oxide, which prevents further surface corrosion and blocks corrosion from spreading. There are over 150 different types of stainless steel, but about 15 are most commonly used. It is available in plates, bars, wires, tubing, sheets, and basically any other form, like normal carbon steel. Over 70% of total production is made of 300 series stainless steel. The higher alloy content of 300 series stainless steels makes them more expensive. Low-carbon versions—for example, 316L or 304L—are used to avoid corrosion problems caused by welding. Grade 316LVM is preferred when biocompatibility is required (such as body implants and piercings).

With respect to micromachining, the primary market is in medical device manufacturing. Most nonimplantable medical devices, such as dental and surgical instruments, are manufactured from commercial-grade stainless steels as they are adequate for use when human body contact is brief. Stainless steels used for implants, however, must be suitable for prolonged contact with the tissue (warm and salty environments). Examples of implantable applications are aneurism clips, bone plates and screws, fixation devices, nails, pins, and joints for ankles, elbows, fingers, knees, hips, shoulders, and wrists. Implants have very specific surface-finish requirements and the surfaces are usually mechanical or electropolished to a high finish. Polished surfaces offer enhance corrosion resistance and offer better biocompatibility. Furthermore, implants are subject to a stringent cleaning and passivation process in order to remove debris and microbiological contamination.

Stainless steels can be processed using a wide range of laser wavelengths, and the laser and optical setup choice depends on the task being performed and the amount of postlaser processing required. Figure 8.1 shows three views (as processed, extended, and close-up) of a 300 μm diameter stainless steel hypo tube cut in a spiral using a fiber laser. The wall thickness is 75 μm and the inner diameter is approximately the diameter of a human hair! Figure 8.2 shows two views of a laser-cut 304 stainless steel solder mask stencil: (a) 200 μm thick and (b) 125 μm thick. These parts were cut using oxygen

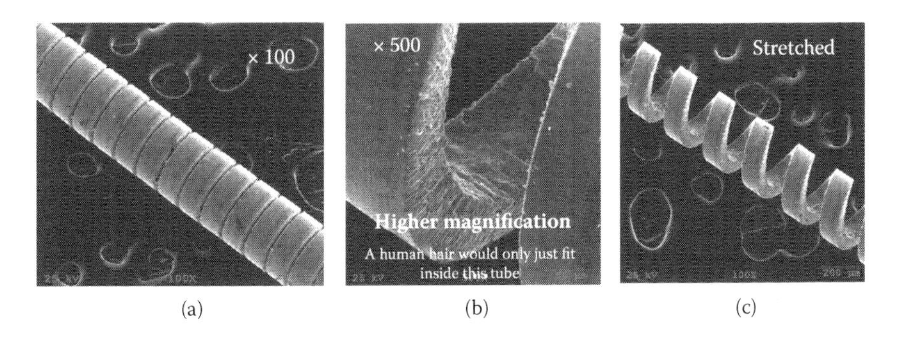

(a) (b) (c)

FIGURE 8.1
(a–c) Stainless steel hypo tube. (Courtesy of LPL Systems.)

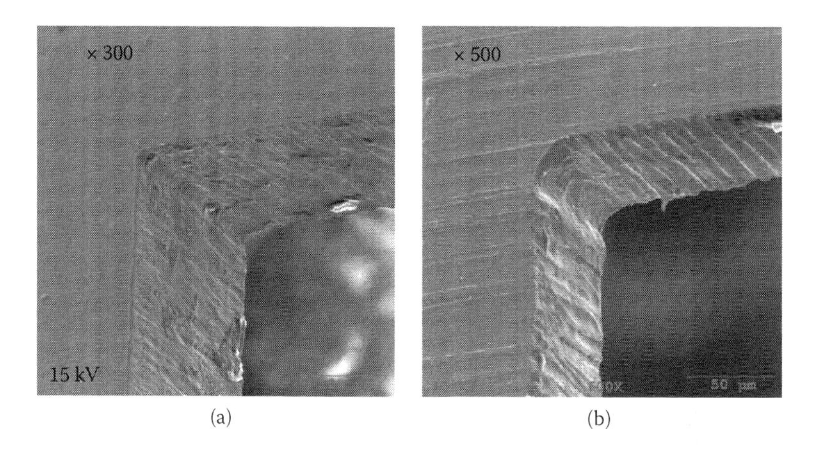

FIGURE 8.2
(a, b) 304 stainless steel stencil edges. (Courtesy of LPL Systems.)

assist. Figure 8.3 shows three different medical parts made using a nanosecond 355 nm laser. Figure 8.4 shows some cuts in 100 μm thick stainless steel using nanosecond, picosecond, and femtosecond lasers.

8.1.2 Copper

Copper (Cu) is an elemental metal with a reddish-orange natural color that will tarnish when exposed to air or moisture. It has high thermal and electrical conductivity—the second highest among pure metals at room temperature. It is also ductile, malleable, and relatively inexpensive compared to other elements in the same grouping of the periodic table, such as silver and gold. The amount of copper in use is increasing rapidly and the available reserves are barely sufficient for current world usage. For example, copper has been in use for at least 10,000 years, but 95% of the current capacity has been mined since 1900. It was discovered thousands of years ago that copper can be alloyed with zinc or tin to make brass or bronze. Recent alloys include beryllium copper (BeCu), phosphor bronze,

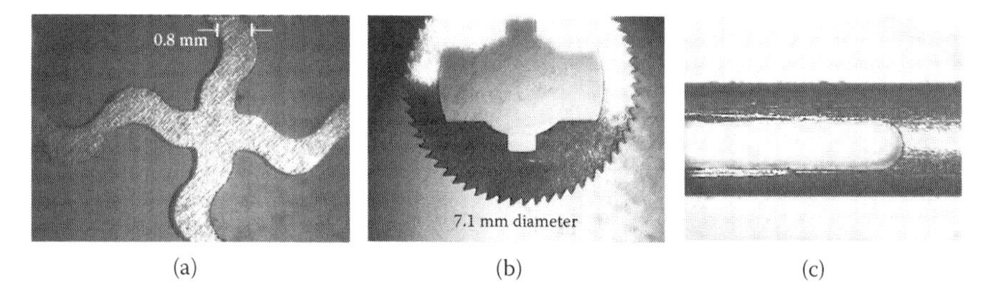

FIGURE 8.3
355 nm laser-cut stainless steel parts: (a) implantable gear; (b) complex shape; (c) eye of a needle.

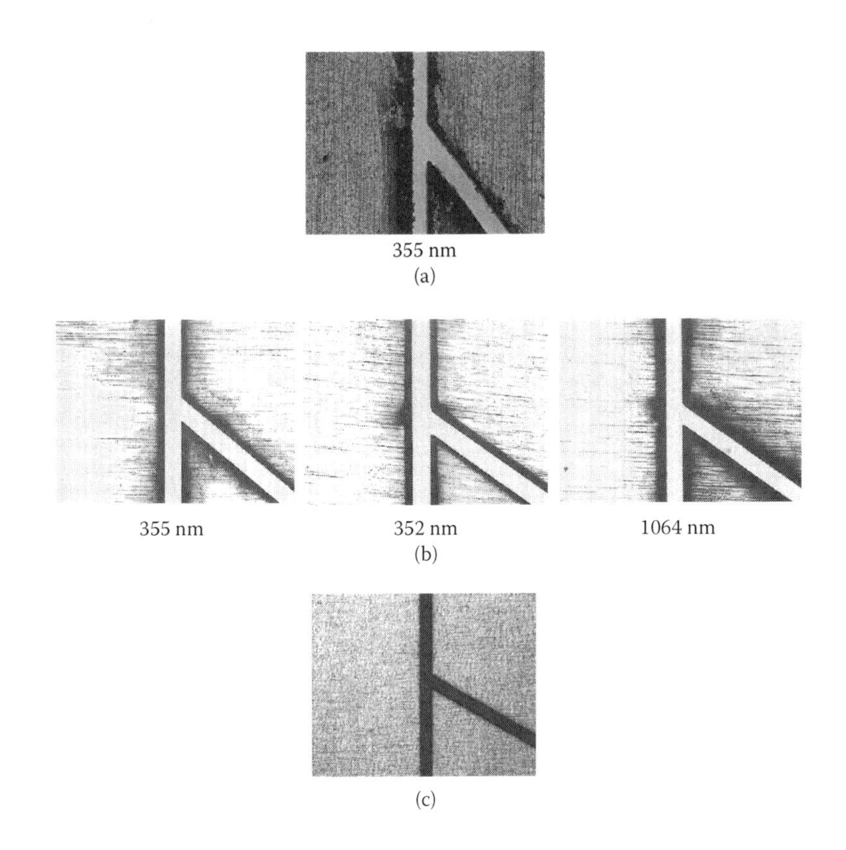

FIGURE 8.4
Cuts in 100 μm thick stainless steel using different lasers: (a) 355 nm, 50 ns; (b) 355, 532, and 1064 nm, 12 ps; (c) 1030 nm, 300 fs.

aluminum bronze, and cupronickel. While pure copper has antimicrobial properties, some of its alloys, like BeCu, are extremely toxic and must be handled with care, especially when laser machining creates debris and vapors.

All of the copper-based metals can be laser machined using ultraviolet (UV) lasers and some near-infrared (IR) lasers. However, all are highly reflective at the 10 μm wavelength of the CO_2 laser, so it is generally not a good candidate for laser processing these metals. Green light at 532 nm wavelength also works quite well on copper-based metals. Copper is used in electrical applications and UV lasers are used to drill microvias in printed circuit boards and flexible circuits. Since the CO_2 laser wavelength is highly reflective, it is used in many hundreds of laser systems to remove dielectric without disturbing the copper conductors. If the copper is thin enough, a highly absorptive "black" layer can be added to the surface of copper sheet. This will initiate a thermal reaction when a CO_2 laser is used and allow vias to be drilled, but this only works if the copper is less than

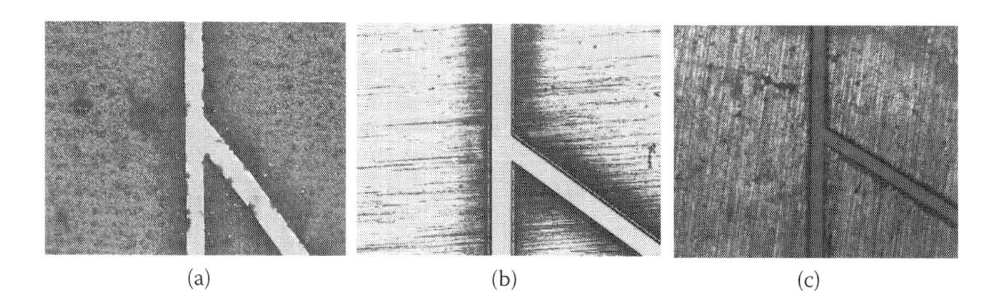

FIGURE 8.5
Cuts in 125 μm thick copper using different lasers: (a) 355 nm, 50 ps; (b) 1064 nm, 12 ps; (c) 1030 nm, 300 fs.

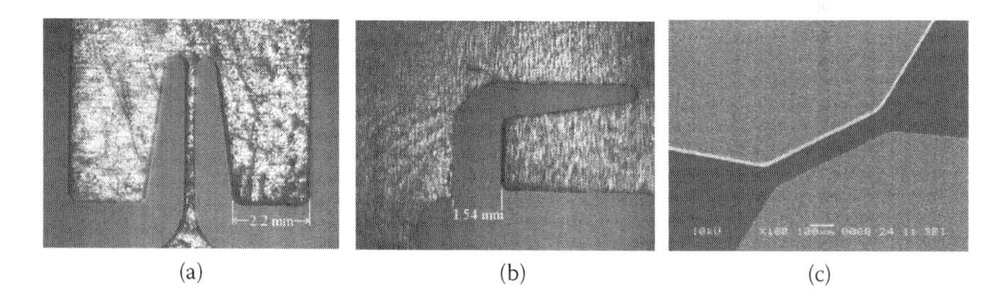

FIGURE 8.6
Laser-cut shapes in copper compounds: (a) BeCu (355 nm, 50 ns), 150 μm thick; (b) phosphor/bronze (355 nm, 50 ns), 300 μm thick; (c) brass (355 nm, 12 ps), 100 μm thick. (Courtesy of Lumera.)

about 15 μm in total thickness. Figure 8.5 shows cuts by three different lasers in 125 μm thick copper. Figure 8.6 shows some shapes cut out of copper alloys.

8.1.3 Molybdenum

Molybdenum (Mo) has the sixth highest melting point of any pure metal and laser machines quite well with near-IR and UV lasers. It also has one of the lowest coefficients of thermal expansion of any metal. It is very stable and does not react with water or oxygen at room temperature. The ability of molybdenum to withstand extreme temperatures without significantly expanding or softening makes it useful in applications that involve intense heat such as the manufacture of armor, aircraft parts, electrical and industrial contacts, motors, and filaments. It is also used in alloys such as stainless steel. Although it does occur naturally in the body, dusts and fumes generated in the machining process can be toxic if ingested or inhaled. Figure 8.7 shows cuts by three different lasers in 50 μm thick Mo. Figure 8.8 shows a serpentine pattern made with a 355 nm, 50 ns pulse-length laser; this photo was taken directly after laser processing with no postlaser cleaning.

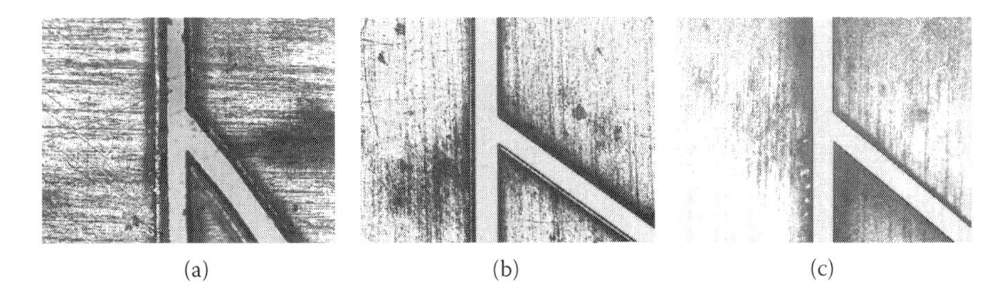

FIGURE 8.7
Cuts in 50 μm thick molybdenum using different lasers: (a) 355 nm, 50 ns; (b) 1064 nm, 12 ps; (c) 355 nm, 12 ps.

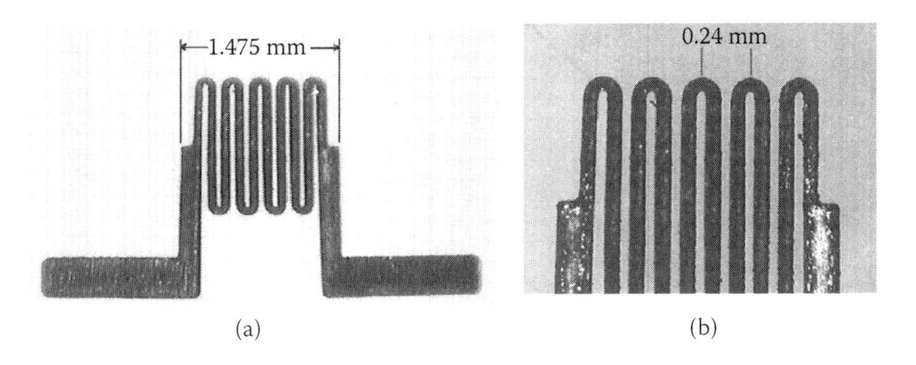

FIGURE 8.8
(a, b) Serpentine pattern in 50 μm thick molybdenum.

8.1.4 Aluminum

Aluminum (Al) is the most abundant metal and the third most abundant element on earth; it is interesting for its low density and ability to resist corrosion due to passivation. Components made from aluminum and its alloys are vital to many industries, including aerospace. It is a good reflector of many wavelengths of light and has very good electrical conductivity. It is the second most widely used metal (behind iron) and has one-third the density and stiffness of steel. It is easily machined, cast, drawn, and extruded. It is almost always alloyed to some extent for industrial usage. It has a very high strength-to-weight ratio. In addition to the aerospace industry, it is used in electronics and CDs, heat sinks, and LED lighting. Extruded aluminum is used in vacuum chucks, bases, gantries, and enclosures in laser micromachining systems. It can be laser machined with near-IR and UV lasers. Figure 8.9 shows cuts by two different lasers in 300 μm thick Al: (a) a 355 nm (50 ns) DPSS laser and (b) a 300 fs laser. Note that in the latter the "chad" was left in the photo.

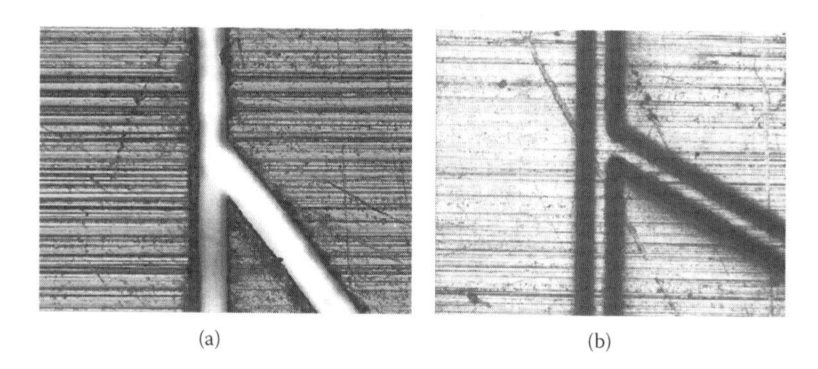

FIGURE 8.9
Cuts in 300 μm thick aluminum using two different lasers: (a) 355 nm, 50 ns; (b) 1030 nm, 300 fs.

8.1.5 Titanium

Titanium (Ti) is another metal that is very corrosion resistant and lightweight. It has a very high melting point and the highest strength-to-weight ratio of any metal. It can be as strong as stainless steel, but 45% lighter. Its most popular compound, titanium dioxide (TiO_2), is used as a white pigment in paints and polymers and its presence makes laser marking very crisp, clean, and indelible. Figure 8.10 shows a mark made with a 355 nm laser onto the surface of a 2 mm diameter catheter containing TiO_2 pigment.

Titanium can also be alloyed with iron, aluminum, molybdenum, and many other elements to produce strong and lightweight metals for jet engines, missiles, spacecraft, medical prostheses, orthopedic implants, dental implants, mobile phones, and dental and endodontic instruments. It is fairly hard and nonmagnetic but a poor conductor of heat and electricity. Therefore, as the material can be distorted easily, traditional machining methods require great care and lasers are a very good choice for precision machining applications. In engine applications, it is used for rotors and compressor blades and in hydraulic system components. The alloy titanium 6AL-4V accounts for almost 50% of all alloys used in aircraft applications. Over two-thirds of all titanium produced is used in aircraft engines and frames,

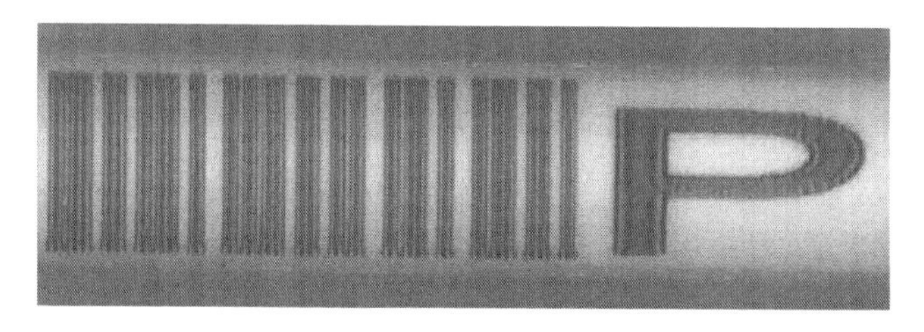

FIGURE 8.10
Laser-marked catheter with TiO_2 pigment.

including such well known classics as the SR-71 Blackbird; Boeing 777, 747, and 737; and Airbus A340, A330, A320, and A380.

Titanium is also used in medical applications. Because it is biocompatible (nontoxic and not rejected by the body), titanium is used in surgical implements and implants, such as hip balls and sockets (joint replacement) that can stay in place for up to 20 years. It is also used in dental implants that can remain in place for over 30 years. It is useful for orthopedic implants, which benefit from titanium's lower modulus of elasticity. This more closely matches that of the bone that such devices are intended to repair. As a result, skeletal loads are more evenly shared between bone and implant, leading to a lower incidence of bone degradation due to stress shielding and bone fractures, which occur at the boundaries of orthopedic implants. Since titanium is nonmagnetic, patients with titanium implants can be safely examined with magnetic resonance imaging.

Figure 8.11 shows several different products laser processed with a 100 W fiber laser. Titanium is also used for surgical instruments where high strength and low weight are desirable. Although titanium is nontoxic, even

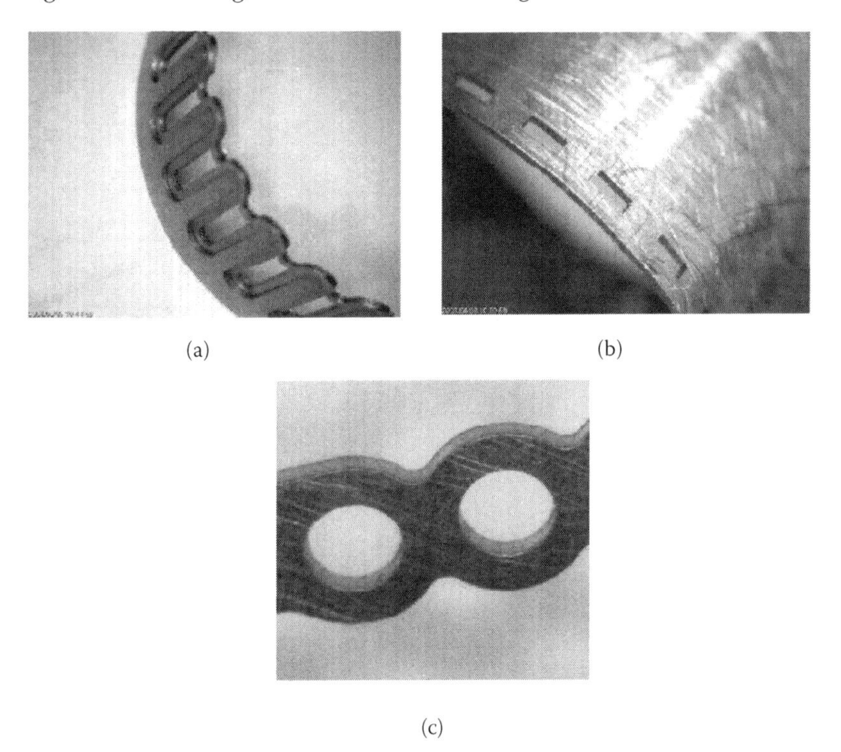

(a) (b) (c)

FIGURE 8.11
Laser processing of titanium with 100 W fiber laser: (Courtesy of LPL Systems.) (a) laser cut heart valves (0.8 mm wall thickness); (b) laser cut features in cylinder (0.8 mm wall thickness); (c) laser cut bone plate.

in large quantities, it can pose an extreme fire and explosive hazard if shavings and dust are heated in air. Therefore, though it welds and cuts beautifully, it only does so in the absence of oxygen.

8.1.6 Nickel and Nitinol

Nickel (Ni) is magnetic and relatively corrosion resistant and is mostly used as an alloy. Nickel and its alloys are frequently used as catalysts, and foams or meshes are used in gas diffusion electrodes and alkaline fuel cells. It is used as a binder and when vapor depositing thin metallic films (discussed later) frequently a very thin nickel layer is placed down first so that the cover metal film has good adhesion. Nickel is also used in electroplating, especially for use with corrosive materials. Dust and fumes from nickel and its compounds can be carcinogenic, so care must be taken when machining. Nickel is used in many products in use in everyday life, including electric guitar strings, magnets, rechargeable batteries, and computer hard drives. Electroformed nickel can be difficult to machine, but near-IR and UV lasers can be used to create very fine, clean features.

Figure 8.12 shows an uncleaned part made from 200 μm thick electroformed nickel with a 355 nm, 50 ns laser showing the delicate serpentine pattern on the sides. Figure 8.13 shows a nozzle and a cone in 1 mm thick chromium nickel, laser processed with a 12 ps, 355 nm laser by using trepanning.

Nitinol is an alloy of nickel and titanium (present in approximately equal amounts) that has two interesting properties: shape memory and superelasticity. It is also highly biocompatible. These properties allow this metal to be used in applications where deformations are induced, and thereafter the material returns to its original shape—for instance in medical applications (primarily stents). Figure 8.14 shows a nitinol stent laser processed with a 10

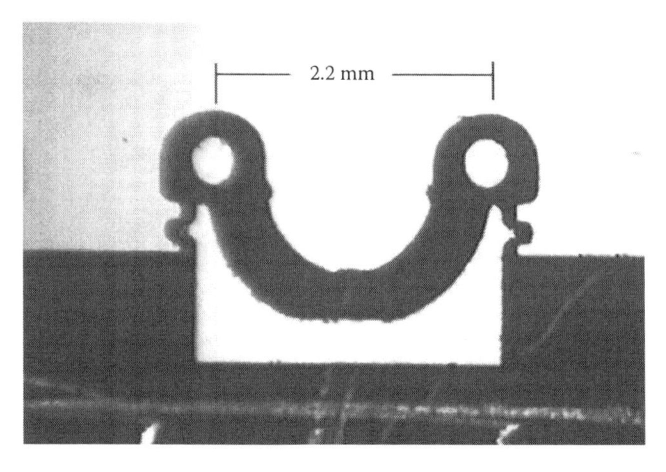

FIGURE 8.12
Laser-patterned electroformed nickel (355 nm, 50 ns).

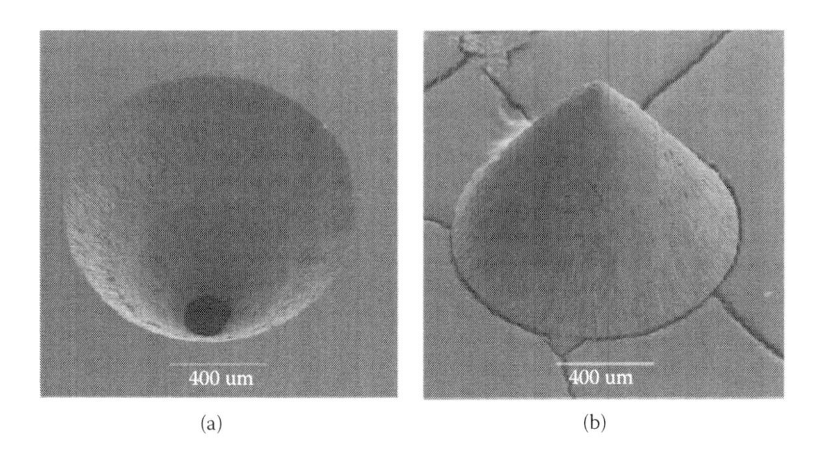

FIGURE 8.13
(a, b) Nozzle and cone shapes in chromium nickel. (Courtesy of Lumera.)

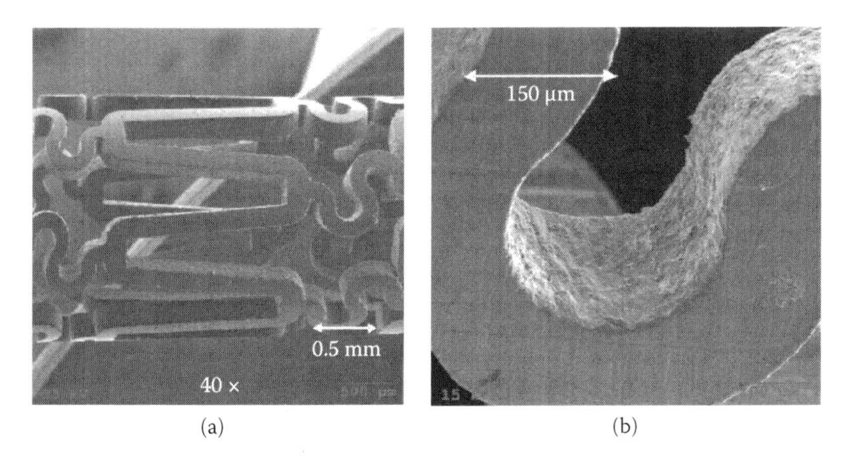

FIGURE 8.14
(a, b) Fiber laser cut nitinol stent. (Courtesy of LPL Systems.)

W fiber laser; the wall thickness is 200 μm. Figure 8.15 shows a nitinol stent laser processed with an 800 fs IR laser, cleaned only in an ultrasonic bath. The wall thickness is 127 μm and the struts are approximately 92 μm.

8.1.7 Thin Metallic Films

Thin films are less than 1 μm in thickness and usually on the order of a few hundred angstroms; they are usually deposited onto a bulk substrate—polymers, ceramics, or even other metals. These thin metallic films react somewhat differently to impinging laser light than the bulk material. In most cases, the fluence required for removal is much lower than that required in

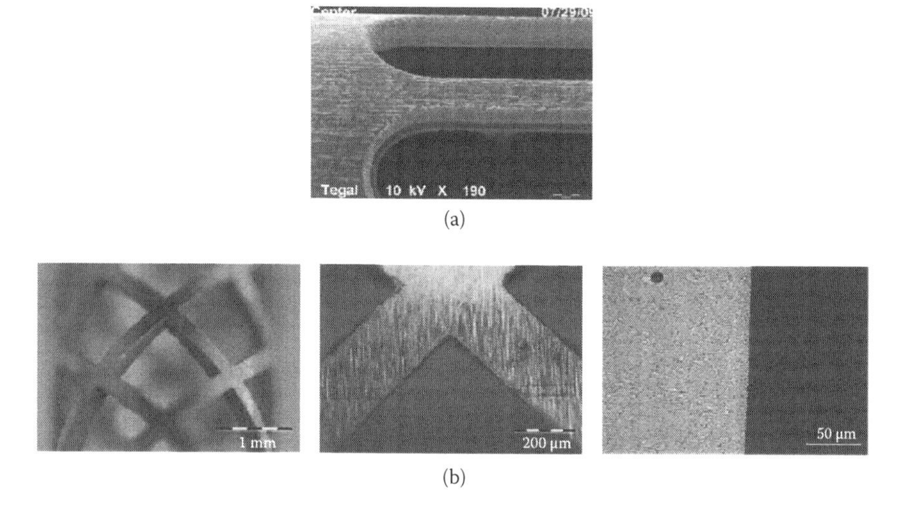

FIGURE 8.15
(a, b) Femtosecond laser-cut nitinol stent. (Courtesy of Raydiance.)

bulk material. This allows very fine patterning on even polymeric substrates with minimal or no damage to the underlying substrate. Material seems to be removed by some combination of ablation and acoustic shock that lifts the film from the surface. Thin films are ideal for patterning with near-IR and UV DPSS lasers.

8.2 Ceramics

Because they are hard materials, ceramics tend to follow the same pattern as seen with metals, in that UV lasers can be used on very thin material. However, as the thickness gets greater than 500 μm, the process becomes slower and perhaps not cost effective. Ceramics in general do not respond well to the 1 μm wavelength; however, the CO_2 laser works quite well for larger features in some ceramics. Hard ceramics and also ceramics in the "green" state can be laser processed.

8.2.1 Alumina

Alumina (Al_2O_3) ceramics have basically the same chemical structure as sapphire and ruby. They are the most widely used ceramic. Beneficial features include high mechanical strength, high electrical insulation, low coefficient of thermal expansion, and high wear and chemical resistance. The 10.6 μm CO_2 laser is particularly well matched to the characteristics of alumina and offers a number of important benefits over mechanical techniques. The low thermal conductivity of alumina means that the laser only produces "local" heating, thus minimizing thermal stress on the part. With 200–400 W CO_2 lasers operating at 10.6 µm wavelength and using coaxial gas assist, exit holes diameters can be achieved in 125 µm thick alumina down to about 50 µm. However, there is a heat-affected zone (HAZ) of up to 100 µm and this laser produces slag on the underside of the cut. Exit hole diameters of about 75 µm can be achieved in 375 µm thick alumina, but the HAZ is about 175 µm.

Cut and drill quality can be improved using an emulsifier on the underside that is washed off after the laser process, but material sometimes needs to be annealed after processing. In principle, CO_2 lasers can handle material up to about 1 mm thick. UV lasers can achieve smaller hole sizes, but in thinner material. The cut quality is much better, but the speed is substantially slower.

Figure 8.16 shows cuts made in 200 µm thick alumina using three different lasers. Figure 8.17 shows 248 nm excimer laser-drilled holes through 1.5 mm of alumina with a nominal 100 µm diameter (110 µm entrance and 90 µm exit). Figure 8.18 shows 10 µm "pyramids" on a ceramic tip. This structuring is similar to the geometry of a meat tenderizer mallet, but on a much smaller scale and in a different material. Figure 8.19(a) shows entrance holes in 400 µm thick alumina made using a 355 nm nanosecond laser. These 10 holes were randomly selected from thousands of generated holes to show consistency in the laser process. The hole diameters range from 205 to 214 µm and the drill time was about 6 s per hole. Figure 8.19(b) shows the same exit holes varying from 162 to 169 µm.

Figure 8.20 shows a very dense array of holes drilled in 200 µm thick alumina with a section of exit and entrance holes shown. Note that the entrance holes have very little material left between the holes, but still maintain their structural rigidity. Figure 8.21 shows excimer laser-marked ceramic capacitors compared to a match head (a) and in high resolution (b). These marks are made with no HAZ or microcracking and at very high speeds since only one pulse is needed to make the mark.

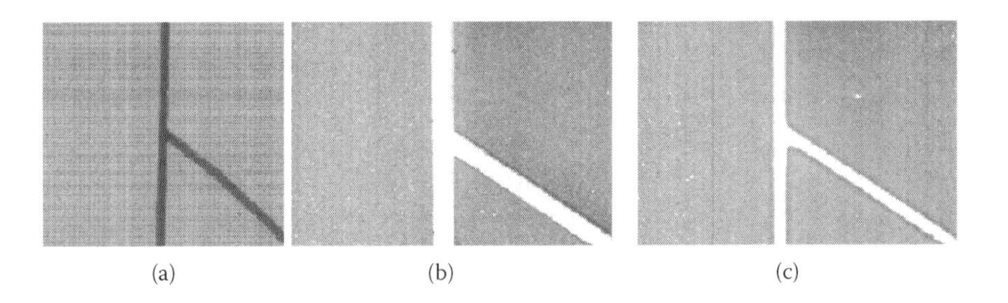

FIGURE 8.16
Cuts in 200 µm thick alumina using three different lasers: (a) 355 nm, 20 ns; (b) 355 nm, 12 ps; (c) 532 nm, 12 ps.

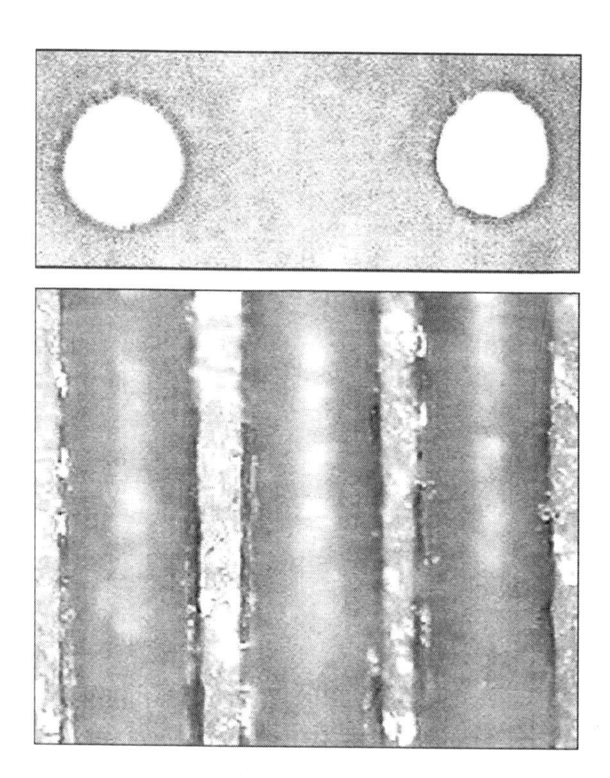

FIGURE 8.17
High-aspect ratio 100 μm diameter holes in 1.5 mm thick alumina.

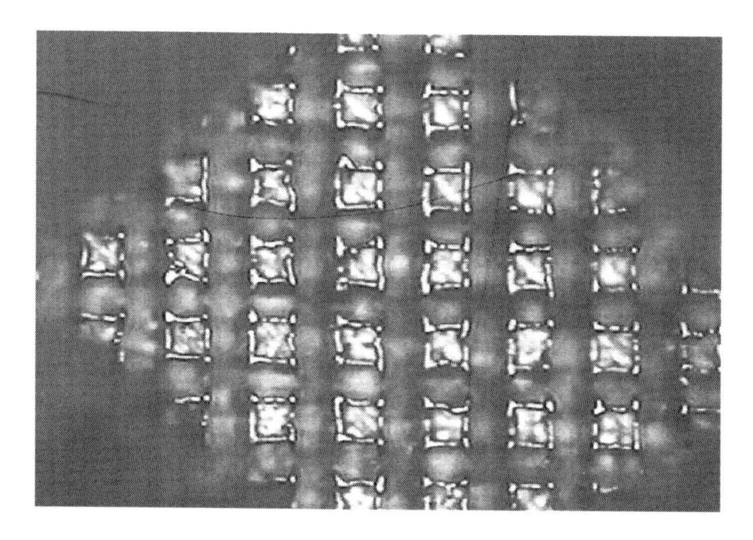

FIGURE 8.18
Pyramids 10 μm on a side on a ceramic tip.

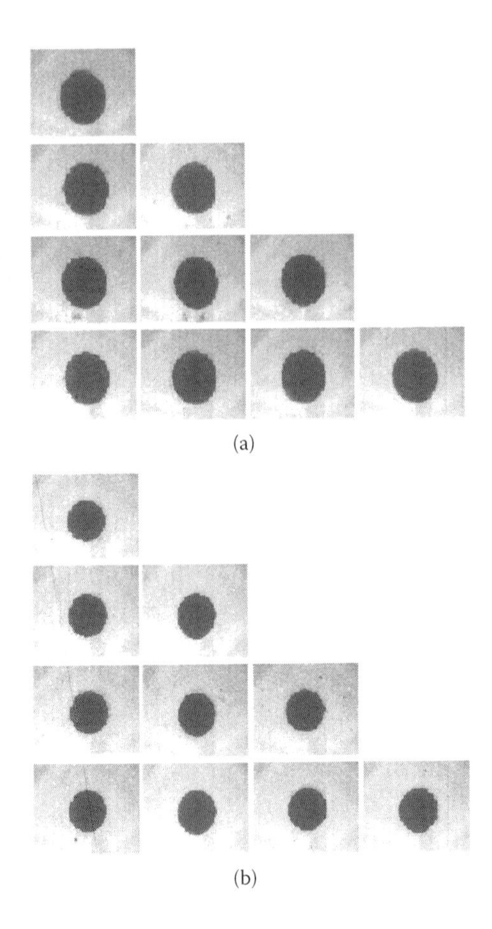

(a)

(b)

FIGURE 8.19
Holes drilled in 400 μm thick alumina: (a) randomly selected entrance holes ranging from 205 to 214 μm; (b) randomly selected exit holes ranging from 162 to 169 μm.

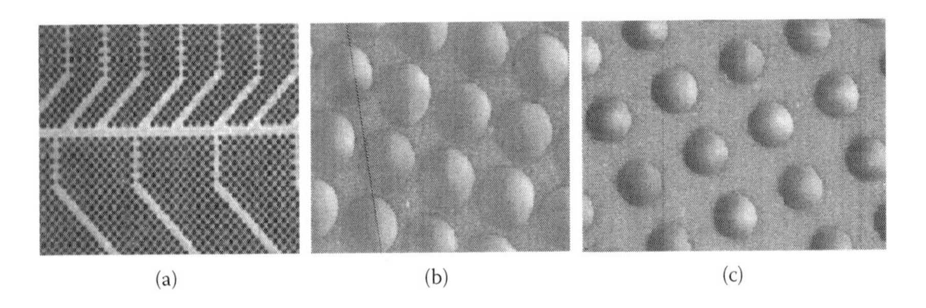

(a) (b) (c)

FIGURE 8.20
(a–c) Dense array of 100 μm holes in 200 μm thick alumina, including entrance and exit holes.

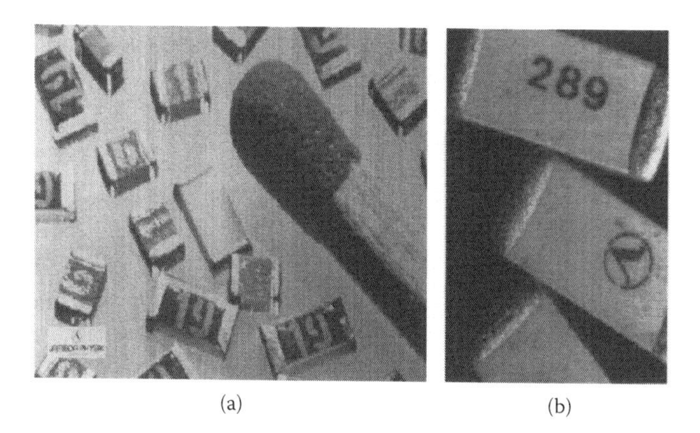

(a) (b)

FIGURE 8.21
(a, b) UV laser-marked ceramic capacitors. (Courtesy of Coherent.)

8.2.2 Silicon Carbide and Silicon Nitride

Silicon carbide (SiC) maintains its strength at very high temperatures and has a very high chemical corrosion resistance compared to other ceramics. It is used in mechanical seal rings and pumps. However, it is difficult to machine using mechanical processes due to its brittleness and hardness. Laser ablation provides a good alternative to traditional machining.

Silicon nitride (Si_3N_4) exceeds most other ceramics in thermal shock resistance. Since it does not deteriorate at high temperatures, it is ideal for use in components used in automotive engines and gas turbines. It can be found in turbocharger rotors, glow plugs for diesel engines, and hot plugs.

8.2.3 Zirconia

Zirconia (>90% ZrO_2) ceramics have high mechanical strength and toughness, even at room temperature, compared to other ceramics. They can also be polished very smooth. Due to these characteristics, they are used in the manufacture of cutting tools (scissors and knives) and in pump applications. Zirconia also has high chemical resistance, low thermal conductivity (<10% that of alumina), and high density. As discussed in Section 7.3.4, zirconia ceramics are used as thermal barrier coatings on jet engine components and there is a constant need to refurbish worn, high-value components.

8.3 Glasses

Glasses are defined as amorphous (noncrystalline) solids that are typically brittle, have low heat conduction, are optically transparent, and exhibit a specific transition when heated to the liquid state. Some of the more common examples include quartz, fused silica, borosilicates (Pyrex™ and BK7), and sapphire. All share common traits in that, because of their optical transparency over a wide range of wavelengths and their low heat conduction properties, they can be difficult to laser machine.

Nevertheless, glasses have attractive properties that make them suitable for many micro- and nanotechnology applications. Very thin sheets of glass (with thicknesses less than 1 mm) are extremely difficult to handle and machine with conventional processes. Using lasers, however, very thin glasses can be machined at high speeds and with high quality and no peripheral damage. CO_2 lasers can be used for glass cutting, scribing, and marking. Methods have been developed that surpass traditional glass cutting/separating methods and result in a cut edge of very high quality. The main advantages of these processes are high precision, no microcracking or chipping, smooth optical quality edges that are resistant to breakage, and cost-effective, one-step operation. Very thin glass can also be processed using a UV or ultrashort pulse laser, both of which can give excellent edge quality.

Figure 8.22 shows approximately 100 μm diameter holes drilled in 500 μm thick glass using a 1 μm wavelength femtosecond laser (left) and a nanosecond laser (right). It is clear that the much shorter pulse length of the femtosecond laser gives much cleaner processing results. Figure 8.23 shows 750 μm thick quartz cut with a 50 W CO_2 laser at 500 mm/min cutting speed. The filament is only 200 μm across at its narrowest point. Figure 8.24 shows 100 μm thick display glass cut with a CO_2 laser.

Other methods for laser processing glass are in-volume selective laser etching (discussed in Chapter 6) and water-jet assisted laser cutting. In traditional water-jet cutting, a high-pressure stream of liquid containing abrasive material is directed onto the part. This abrasive material does the etching

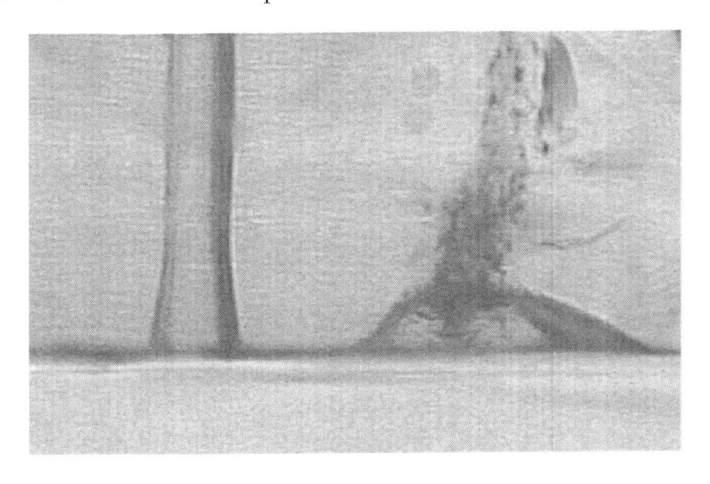

FIGURE 8.22
100 μm diameter holes in 500 μm thick glass. (Courtesy of Clark MXR.)

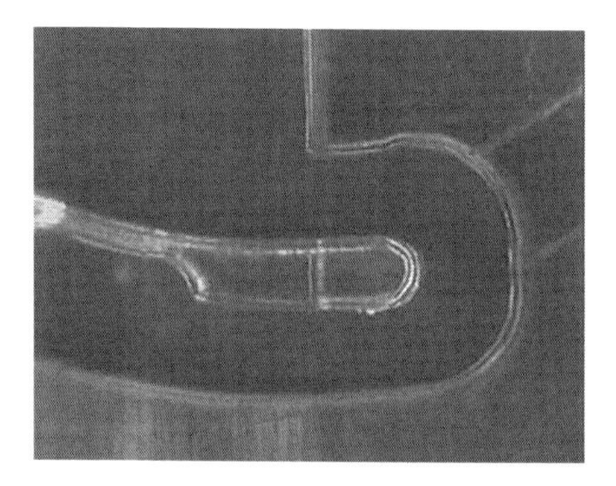

FIGURE 8.23
CO_2 cut quartz. (Courtesy of Synrad.)

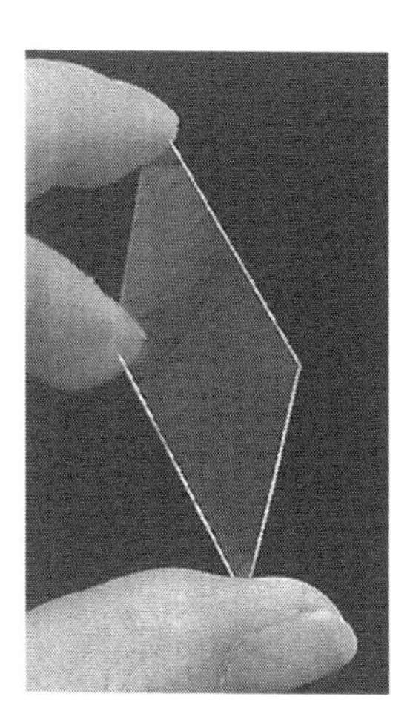

FIGURE 8.24
Thin glass sheet cut with CO_2 laser. (Courtesy of Synrad.)

while the liquid cools the process and keeps the amount of debris low. In the laser-jet assisted cutting process, a laser replaces the abrasive and is focused coaxially with the water-jet. This combination procedure produces very high-quality cuts at high speeds with no microcracks. Both methods, however, get the parts wet and dirty, so a subsequent cleaning step is needed. Also, in some cases the addition of a liquid during processing is undesirable, especially, for instance, when the glass has other features already present on it (like electrical circuits).

Lasers can also be used to shape the tips of glass fibers, creating lenses or other shapes. Lenses produced on the ends of optical fibers, for instance, have a host of uses based around their ability to improve and control the coupling of light into and out of the fiber. These lenses are used extensively in fiber optic telecommunications, and they also provide efficient and flexible optical probes in other applications, such as biomedical gel scanning, silicon-wafer fluorescence spectroscopy, and on-die VCSEL (vertical cavity surface-emitting laser) characterization. Short-pulse lasers are very useful on glasses due to their small HAZ.

Figure 8.25 shows deep mesas in sapphire, 90 μm on a side, processed with a 355 nm picosecond laser. Figure 8.26 shows a pattern of stepped concentric rings in a sapphire wafer. The rings are 100 μm wide and the step height is 10 μm; they were made using a picosecond laser operating at high repetition rate.

FIGURE 8.25
Picosecond laser processing of mesas in sapphire. (Courtesy of Lumera.)

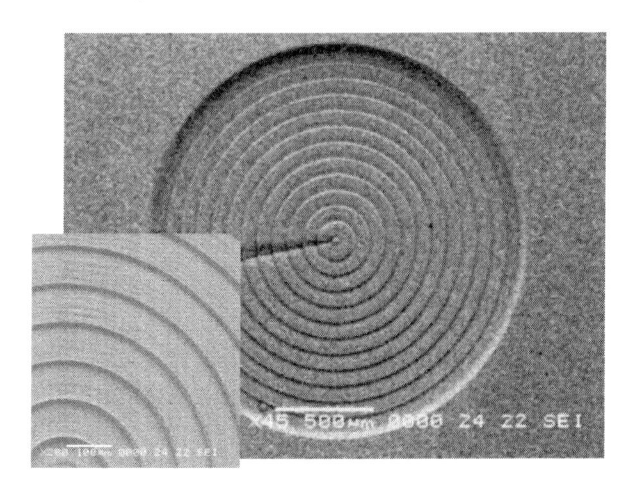

FIGURE 8.26
Picosecond laser processing of stepped concentric rings in sapphire. (Courtesy of Lumera.)

Finally, decorative sculptures can be made using lasers that focus into the transparent material and create microbursts inside the bulk that, when done with three-dimensional software, result in a visible sculpture. These have been made in very large formats; life-size sculptures and small pieces are sold in novelty shops for only a few dollars each.

8.4 Silicon and Gallium Arsenide

Silicon (Si) is the second most abundant element on earth and is used in a variety of different applications, primarily in the semiconductor industry for integrated circuits, transistors, and microchips. In recent years, silicon has also been used extensively in photovoltaic applications. Silicon can be cut, drilled, or scribed; in addition, surface modifications can be made—for instance, to increase surface area. When working with polished wafers, a sacrificial layer of photoresist can be applied that, when washed off, carries away debris generated by the laser process. A commonly performed operation on discs is resizing to smaller sizes. Wafers up to 300 mm in diameter can be manufactured in bulk; however, smaller companies or companies with older equipment frequently cannot handle wafers this big, so it is cost effective to use a laser to resize.

The 1 μm wavelength works very well on Si for most purposes, and the use of an assist gas gives the best cutting and cleanest edges. The key in using near-IR lasers is to reduce microcracking down to less than 5 μm as required by the electronics industry; new processes being developed, such as "stealth dicing," address this issue. Wafer thicknesses of up to 1 mm can be processed, but typical wafer thickness is 500 μm or less. The thinner the silicon, the cleaner and faster will be the laser process. For front-to-back contact via holes, percussion drilling is used for smaller diameter holes and trepanning is used for larger diameter holes. Microelectromechanical systems devices are also made in silicon using lasers with pockets, channels, and through-features down to 10 μm feature size.

Figure 8.27 shows cuts made in 500 μm thick Si using a 100 W, continuous wave (CW) fiber laser with gas assist (a) and a 355 nm nanosecond laser (b). The cut speed of the fiber laser is over 300 mm/s, while the 355 nm UV laser cut speed is about 10 mm/s; however, the quality of the UV cut is better even without gas assist. Figure 8.28(a) shows a cut made using a femtosecond laser, and the chad is shown in Figure 8.28(b). Note that the chad is only about 50 μm wide and still maintains its mechanical properties without any compromise from HAZ.

Gallium arsenide (GaAs) is a II/V semiconductor that is used in microwave circuits, IR light-emitting diodes, laser diodes, solar cells, and optical windows. It has some properties that are superior to those of silicon, such

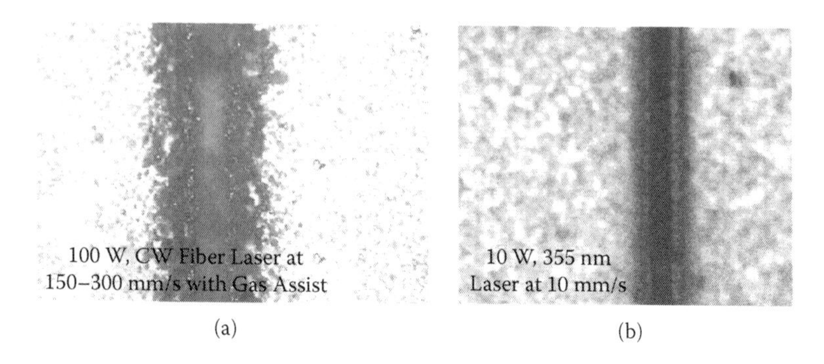

FIGURE 8.27
Cuts with two lasers in 500 μm thick silicon: (a) 100 W CW fiber laser with gas assist; (b) 10 W, 355 nm nanosecond laser.

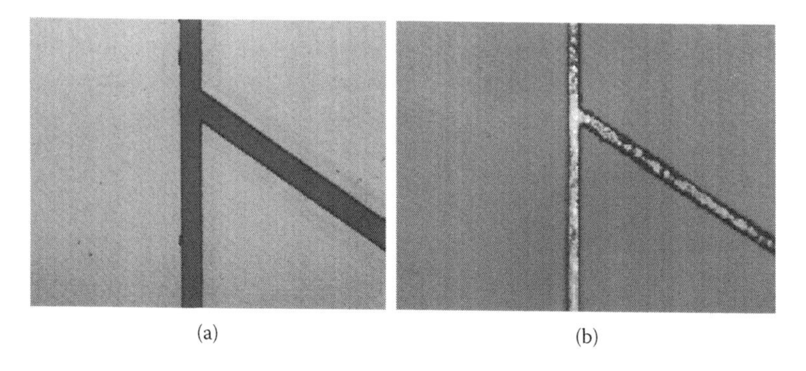

FIGURE 8.28
Femtosecond laser cuts in Si: (a) cut area with chad removed; (b) chad.

as a higher saturated electron velocity and higher electron mobility. High-speed transistors made from GaAs operate at frequencies in excess of 250 GHz. Another advantage is that it has a direct band gap, which means that it can be used as an efficient light emitter and for high-efficiency (and high-cost!) solar cells. Due to its arsenic content, it is potentially hazardous to work with, especially if vaporized. The crystal is stable enough not to cause any concerns, but fine powders and vapors are considered carcinogenic and great care should be taken when laser processing this material.

8.5 Polymers

Polymers are especially conducive to the use of short-wavelength and short-pulse lasers, although many respond well to far-IR lasers as long as the required feature sizes are not particularly small and some HAZ can be accepted. CO_2 lasers work especially well on acrylics and leave a very smooth and almost polished edge when gas assist is used in cutting. The absorption of UV photons is especially enhanced when π-electrons are present in the material. Near-IR lasers are generally not as useful on polymers as either longer or shorter wavelength lasers.

8.5.1 Parylene™—Poly(P-Xylylene) Polymers

Parylene is a vapor-deposited polymer used as a chemical, moisture, and dielectric barrier. It is frequently used as a complete covering over complicated topographies and in electronics for space travel and military applications. In the medical market, its uses are diverse and widespread. It offers a level of protection to critical components that is unequaled by any other coating. Parylene is biocompatible, pinhole free, and extremely thin, and it protects against the effects of fluids and solvents. It is also nontoxic and conforms precisely to substrate contours. It is used in applications where dry lubricity is important, such as catheters, syringes, and implants. Finally, it adds minimal weight and volume to the finished product.

On the "negative" side, parylene is difficult to remove cleanly and precisely once it is applied; traditional methods include thermal (burning the material away with a flame or hot iron), mechanical (picks, sanding, scraping) or abrasion using microblasting of fine grit. All of these methods have obvious drawbacks. Fortunately, UV lasers interact extremely well with parylene and can be used to remove the material cleanly down to the substrate surface. This work as been performed with excimer lasers and also with 355 and 266 nm DPSS lasers. A very gentle cleaning may be necessary after laser processing in order to clean up minimal debris or possible carbonization, but the resulting surface is ready for the next step in the processing line. Figure 8.29 shows laser removal of parylene from coated IC pads.

8.5.2 Teflons™—Polytetrafluoroethylene and Nylons™ (Polyamides)

Polytetrafluoroethylene (PTFE) is a high molecular weight compound consisting of only fluorine and saturated carbon atoms. In industry it is used as a lubricant and a dielectric coating. There are no π-electrons; therefore, it does not absorb UV photons well unless very short vacuum wavelengths are used. CO_2 lasers can be used to remove or pattern PTFE, but the resulting cut quality shows a heat-affected zone. By adding adulterants, UV lasers can be used since the absorption is enhanced by the impurities. Picosecond and femtosecond lasers have also been shown useful in PTFE drilling, cutting, and patterning. Figure 8.30 shows an edge at high magnification of a 1 mm diameter hole cut in 200 μm thick PTFE using a picosecond laser. Similarly, Nylons™ are usually difficult to laser machine, but they do respond somewhat to 248 and 193 nm excimer laser irradiation and ultrashort pulse (USP) lasers.

FIGURE 8.29
Laser removal of parylene from coated IC pads. (Courtesy of Lizotte Tactical, LLC.)

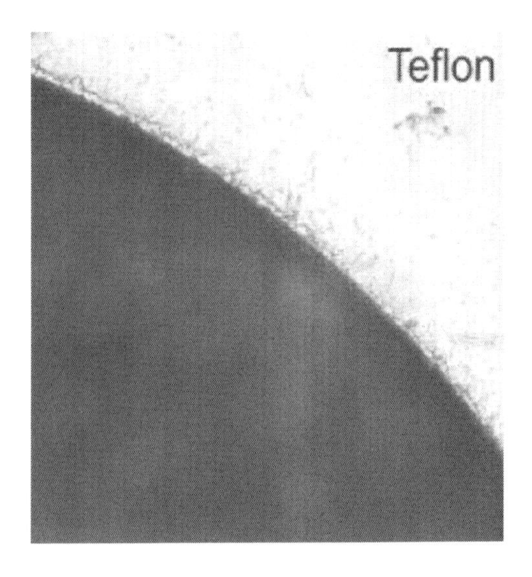

FIGURE 8.30
Edge of 1 mm diameter hole cut in 200 μm thick PTFE using picosecond laser. (Courtesy of Lumera.)

8.5.3 Silicone

Silicones are inert synthetic compounds with a variety of forms and uses. Silicone rubber is inert and heat and moisture resistant; it is used as a sealant and in medical applications and insulation. Silicones are polymers that include silicon together with carbon and other elements. They are good electrical insulators with good thermal stability, low chemical reactivity, moisture resistance, and low toxicity and they do not "stick." Thin rubber components such as gaskets and valves can be easily and quickly cut using 10.6 μm CO_2 lasers; although the cutting is a thermal process, material breakdown occurs at low laser power, so cutting speeds are rapid. Laser cutting imparts no mechanical force on the product, which is easily deformed by conventional contact cutting methods.

8.5.4 Kapton™ and Upilex™ (Polyimides)

Polyimide is a material that remains stable over a very wide temperature range. It is resistant to chemicals; has useful electrical, mechanical, and thermal properties; and is used in a very wide array of applications. In the aerospace industry it is used primarily as an insulator and for wire coating. Due to its ability to withstand flexing without developing cracks or tears, it is used in flexible circuits, diaphragms, and other components that must move constantly under high pressure while performing for millions of cycles. Flexible circuits take advantage of its ability to withstand copper etching and it is used heavily in the electronics industry, especially as the market moves to smaller, lighter, and faster components. As an insulator, its ability to withstand extreme environments makes it ideal for use in the motor and generator industry. Within the photovoltaic market, it is being used in thin film amorphous silicon modules and copper indium gallium selenide applications.

As has been shown in previous figures, Kapton™ can be laser ablated with almost any wavelength laser with varying degrees of process speed and cut quality. It is a material of choice for laser ablation and its properties as related to laser micromachining are very well known and understood. It is supplied as sheets, rolls, and conformal coatings in thickness up to 175 μm. Thicker material is available in laminated form, and insulating woven thread material is also available. It can be supplied in conjunction with pressure- and heat-sensitive adhesives, sometimes with a Mylar™ peel layer. Using UV lasers, it is possible to score through the release layer and adhesive and stop at the polyimide substrate to pattern circuits. Simple pressure-sensitive tape is available as well.

An interesting note about Kapton™ is that it contains impurities that can cause problems when laser drilling small, highly tapered holes such as those used in ink-jet cartridges or some flow orifices. Liquid flow applications that require high degrees of taper are drilled at low fluence., While this is sufficient to ablate the polyimide, it is not sufficient to ablate the impurities,

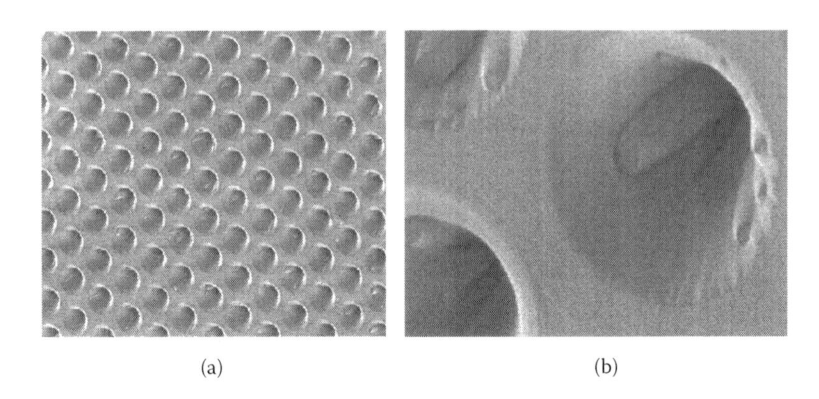

(a) (b)

FIGURE 8.31
(a, b) High-resolution image showing cones in Kapton™.

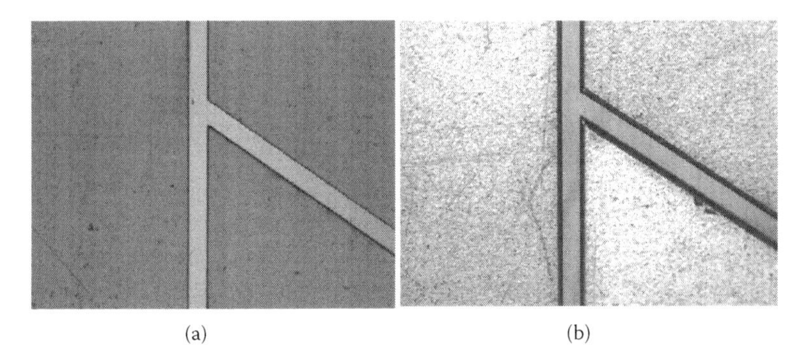

(a) (b)

FIGURE 8.32
Femtosecond laser processing: (a) 50 μm thick Kapton™; (b) 75 μm thick Mylar™.

which can remain inside the holes as "cones" that impede fluid flow and alter the shape and size of the exit hole. Figure 8.31 shows a scanning electron microscope image of such a laser-drilled hole with the cone clearly visible as an obstruction. Figure 8.32 shows (a) Kapton™ and (b) Mylar™ cut with a femtosecond laser. Compare these photos to previous ones and note the comparatively clean edges and cut quality in both materials compared to the cuts made with a longer pulse length 1 μm laser.

8.5.5 Mylar™

Mylar™ (polyethylene terephthalate, or PET) is a polyester film made from stretched PET and is useful for its high tensile strength, chemical and dimensional stability, transparency, gas barrier properties, and electrical insulation. It is also quite inexpensive compared to most other sheet materials. As was shown in earlier figures, it responds well to laser ablation, especially in the UV and it can be cut, drilled, and marked readily. It can be used as a masking material on vacuum chucks and as a template for laser ablation of complex parts: Mylar™ templates are made and then laid over the real parts to check for dimensional tolerances before processing the real parts. This results in risk reduction when complex, high-value parts are being processed—especially if no setup parts are available. It can be metalized and the thin coatings can then be laser patterned with minimal damage to the substrate. It can also be used as a gasket or diaphragm material in manufacturing of disposable medical devices.

8.6 Diamond

Diamonds are the hardest natural material and therefore difficult to machine by traditional methods. Since diamonds are simply tetrahedral carbon matrices, they absorb several wavelengths of light quite readily, and lasers play a big role in marking and removing occlusions from gemstones and also in patterning CVD (chemical vapor deposited) diamond films. The advantages of CVD diamond include the ability to grow diamond over large areas, the ability to grow on a substrate, and the control over the exact properties of the diamond produced. Growth areas of greater than 15 cm diameter have been achieved and much larger areas are likely. The 1 μm wavelength couples very well and can be used to cut or drill diamond with very fast speeds and very good edge quality.

For gemstones and machine tools, this laser is preferred because of the low cost of processing. In the case of machine tools, slight graphitization on the edges is tolerable since a final dressing step is usually performed. However, when the ultimate in edge quality is needed, a UV laser and/or USP laser can be used to cut CVD diamond to exact shapes and sizes with no degradation in edge quality, graphitization, chipping, or microcracking. An example is in the manufacturing of heat sinks. One of the challenges to making microelectronics smaller and faster is heat management and CVD diamond is being used to address this problem as diamond, in addition to being the hardest substance, also has the best thermal conductivity.

Figure 8.33 shows three views of laser-cut diamond. Figure 8.33(a) was cut using a 20 W, q-switched fiber laser; the cut speed is quite high and the edge quality is very good, although there is a bit of carbonization on the edges. Figure 8.33(b) shows a pristine edge cut with a 248 nm excimer laser that shows no signs of microcracking, graphitization, or chipping. Figure 8.33(c) was cut using a 532 nm Nd:YLF laser and again shows a very clean edge with no obvious HAZ.

(a)

(b)

(c)

FIGURE 8.33
Three cuts in CVD diamond: (a) 20 W q-switched fiber laser; (b) 248 nm excimer laser; (c) 524 nm doubled Nd:YLF laser.

9

Metrology and Cleaning

9.1 Metrology

Micromachining, as the name implies, involves making very small features. Parts are generally made according to a drawing file. The normally accepted file format is DXF. AutoCAD DXF (drawing exchange format) is a CAD data file format developed for enabling data interoperability between different CAD programs. Since some CAD programs do not support this file type, especially for three-dimensional processing, many applications use the DWG or some other format such as HPGL, Excellon II drill file, or Gerber.

Regardless of the file type, a drawing must include all relevant dimensions, tolerances, scale, units (English or metric; see Table 9.1), material type, and thickness at a minimum, and it should also include a complete legend (see Figure 9.1) and any other notes needed to clarify the job. The task is to use the laser to make the part as shown in the drawing and to fall within acceptable tolerances on dimensions, cost, and part cleanliness. In order to do this, the parts need to be seen and measured.

Optical comparators are devices that apply the principles of optics to the inspection of manufactured parts. In a comparator, the magnified silhouette of a part is projected upon the screen, and the dimensions and geometry of the part are measured against templates on that screen or by using movable stages and crosshairs. These devices give a great overall view of parts at fairly high magnification (up to 100×), have a workable depth of field, and are found in most machine shops. Large comparators can handle part loads up to 100 lb with several feet of travel.

Stereoscopes use optics to create or enhance the illusion of depth in an image by presenting two offset images separately to the left and right eye of the viewer. Both of these two-dimensional offset images are then combined in the brain to give the perception of three-dimensional depth. Stereoscopes are perfect for viewing parts with up to 100× magnification. Parts can be viewed through eyepieces (binocular) or by using a camera (trinocular) to cast the image on

TABLE 9.1
English to Metric Conversions

1″	One inch		2.54 cm	25.4 mm
0.001″	25.4 microns	One "thousandth"	1 "mil"	0.0254 mm
0.010″	254 microns	Ten "thousandth"	10 mils	0.254 mm
0.040″	1016 microns	Forty "thousandth"	40 mils	1.016 mm

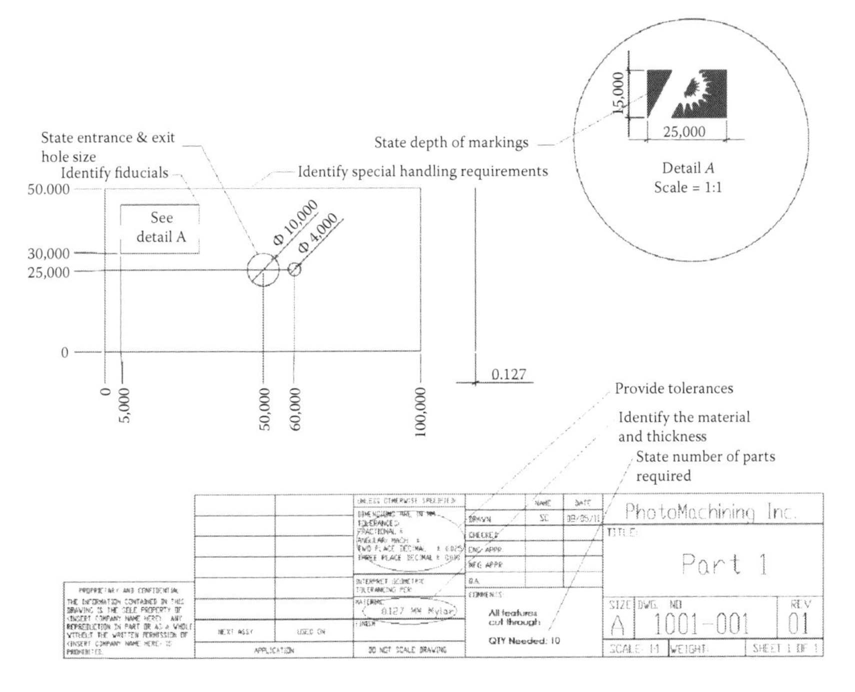

FIGURE 9.1
Drawing file with complete information.

a magnified viewing screen. Because the magnification is low, the field of view of a good stereoscope is larger than with a higher magnification inspection scope.

Inspection microscopes have much higher imaging capabilities than stereoscopes, but with a more limited field of view and depth of focus. Up to 1,000× magnifications are possible.

Measuring microscopes have additional crosshair generators and either manual or motorized x/y/z capabilities, so extremely accurate measurements can be made. Encoders of 0.1 μm are typical on the three axes, so parts can be accurately measured down to a couple of microns.

Coordinate measuring systems are basically automated inspection scopes. These systems incorporate vision systems that can be trained to recognize fiducials or other part features. They are especially useful when multiple repeat measurements are required. For micromachining, these tools use light, but for larger three-dimensional parts, probes can also be used. Optical coordinate measuring systems are accurate down to a few microns.

Scanning electron microscopes (SEMs) are a type of electron microscope that images a sample by scanning it with a high-energy beam of electrons. The electrons interact with atoms that make up the sample, producing signals that contain information about the sample's surface topography, composition, and other properties such as electrical conductivity. The signals result from interactions of the electron beam with atoms at or near the surface of the sample. In the standard detection mode, secondary electron imaging (SEI), the SEM can produce very high-resolution images of a sample surface, revealing details less than 1 nm in size. Due to the very narrow electron beam, SEM micrographs have a large depth of field yielding a characteristic three-dimensional appearance useful for understanding surface structure. A wide range of magnifications is possible, from about 10× (about equivalent to that of a powerful hand-lens) to more than 500,000× (about 250× the magnification limit of the best optical microscopes). Samples must be of an appropriate size to fit in the specimen chamber and are generally mounted rigidly on a specimen holder. For conventional imaging in the SEM, specimens must be electrically conductive, at least at the surface. Metal objects require little special preparation for SEM viewing except for cleaning and mounting. Nonconductive specimens tend to charge when scanned by the electron beam, and, especially in the secondary electron imaging mode, this causes scanning faults and other image artifacts. Parts are therefore usually coated with an ultrathin coating of electrically conducting material (commonly gold) deposited on the sample by low-vacuum sputter coating or high-vacuum evaporation.

Surface profilometers are instruments used to measure a surface in order to quantify its roughness. There are both contact and noncontact types. A typical diamond stylus profilometer can measure vertical features ranging in height from 10 nm to 1 μm. The height position of the diamond stylus generates an analog signal that is converted into a digital signal and then stored, analyzed, and displayed. The radius of the diamond stylus ranges from 20 nm to 25 μm, and the horizontal resolution is controlled by the scan speed and data signal sampling rate. The stylus tracking force can range from less than 1 to 50 mg. An optical profilometer is a noncontact method for providing the same information as a stylus-based profilometer. Different techniques are currently being employed but laser triangulation is probably the most common. Vertical resolution is usually in the nanometer level while lateral resolution is usually less, limited by the wavelength of the light. Because the noncontact profilometer does not touch the surface, the scan speeds are dictated by the light reflected from the surface and the speed of the acquisition electronics. Optical profilometers do not touch the surface and therefore it cannot be damaged by surface wear or careless operators. The spot size, or lateral resolution, of optical methods ranges from a few microns down to the submicron level.

One final note regards what is known as *dimensional analysis.* When making calculations, *always* carry forward units and make sure the final units match your expected answer. Never simply write down numbers as this is a sure way to make a mistake. Dimensional analysis is a tool to check relations among physical quantities using their dimensions. If the dimensions are wrong at the end of a calculation, a mistake has been made somewhere. There are a few basic physical dimensions, such as time, mass, length, temperature, and electric charge. All other dimensions are combinations of these—for example, velocity, which is length per unit time and can be expressed in units of meters per second, feet per second, miles per hour, etc.

Dimensional analysis is based on the fact that a physical law must be unit independent; the corollary is that the units must match appropriately and must be interchangeable in different systems. Some things are dimensionless (e.g., numerical aperture and index of refraction). For instance, whether length is expressed in units of microns, millimeters, inches, mils, feet, or any other unit, the actual physical length itself is unchanged. Therefore, any equation must have the same units on both sides of the equation. Carrying and checking dimensions is a straightforward way of minimizing the potential for error in mathematical and physical calculations.

9.2 Postlaser Cleaning

Cleanliness of laser processed parts is of paramount importance, especially for medical, semiconductor, and aerospace components. Even when performing "clean" laser processing at UV wavelengths and with short pulses, some debris is usually formed during the laser process. The best possible approach is to minimize the amount of debris generated during this process. Even so, some amount of postlaser cleaning usually needs to be done, whether it is a simple wash or a more complicated procedure. Some postlaser processing cleaning techniques are discussed in the following subsections.

Note that cleaning parts is a somewhat risky business—especially for a contract manufacturer because it may not always be apparent that the post-laser cleaning techniques will not cause bigger problems than the debris. Normally, contract manufacturers will not clean laser-processed parts unless there is a clear understanding with the customer concerning the techniques used so that there are no incompatibilities. Postlaser cleaning techniques must be well documented, and established procedures must be followed rigorously to assure the proper end product.

9.2.1 Physical Scrub

The simplest and most economical method of part cleaning is a gentle physical scrub. This is a contact method, so extreme care needs to be taken regarding the scrub solution (if any is used), the scrub tool, aggressiveness of the scrub, and time of scrub. Q-tips, clean room wipes, or soft brushes are normally used on the small and delicate parts manufactured using laser micromachining; acetone, methanol, ethanol, or water (perhaps with a cleaning surfactant) is used as the solvent. Sometimes even the most common and inexpensive things can be used and give excellent cleaning results. For instance, a pencil eraser is quite handy for cleaning carbon and debris from contact pads on printed circuit boards. Another simple technique is to use pieces of a Styrofoam cup and gently rub. The cost for using these techniques (aside from man-hours) is low.

9.2.2 Chemical Bath

An ultrasonic cleaner is a cleaning device that uses ultrasound (usually from 15 to 400 kHz) that is often employed for cleaning delicate items like jewelry, optics, precious metals, surgical instruments, and electronics. In an ultrasonic cleaner, the object to be cleaned is placed in a chamber containing a suitable ultrasound conducting fluid (an aqueous or organic solvent, depending on the application). In aqueous cleaners, the chemical added is a surfactant that breaks down the surface tension of the water base. An ultrasound generating transducer is built into the chamber or may be lowered into the fluid. It is electronically activated to produce ultrasonic waves in the fluid. The main mechanism of cleaning action is by energy released from the creation and collapse of microscopic cavitation bubbles, which break up and lift off dirt and contaminants from the surface to be cleaned. The often harsh chemicals traditionally used as cleaners in many industries can be reduced or eliminated with the introduction of ultrasonic technology.

Devices for home and hobby use are readily available and cost as little as $50. However, typical units for production are priced in the range of a few hundred to a few thousand dollars and the operating costs are very low, assuming that there is no need for exotic liquids.

9.2.3 Electropolishing

A metal workpiece is immersed in a temperature-controlled bath of electrolyte and connected to the positive terminal (anode) of a DC power supply; the negative terminal is attached to an auxiliary electrode (cathode). A current passes from the anode, where metal on the surface is oxidized and dissolved in the electrolyte. At the cathode, a reduction reaction (normally hydrogen evolution) takes place. Electrolytes used for electropolishing are most often concentrated acid solutions with a high viscosity, such as mixtures of sulfuric acid and phosphoric acid.

To perform electropolishing of a rough metal surface, the protruding parts of a surface profile must dissolve faster than the recesses. This behavior (referred to as anodic leveling) is achieved by applying a specific electrochemical condition. In electropolishing, there is a significant preference to the removal of any high spots on the metal surface. This means that the dimensions of the high spots are changed drastically, while the dimensions of the lower spots are changed very little. This creates a smoothing effect to the metal surface.

It also means that by nature of the process, the total amount of dimensional change required to obtain the polish effect is very small. Dimensional reduction of the workpiece is on the order of 2.5 10,000th of an inch/0.00025 in. Electropolishing has many applications in the metal finishing industry because of its simplicity and it can be applied to objects of complex shape. Typical examples are electropolished stainless steel surgical devices.

9.2.4 Plasma Etch

Using plasma eliminates wet processing, reduces chemical disposal cost, and reduces water usage and treatment costs. Labor costs are lowered as well since there are no baths to maintain. With plasma etching, the panels are placed in a vacuum chamber, and gas is introduced and converted to reactive plasma by a power supply. The plasma reacts at the panel surface and volatile by-products (resin smear) are removed by the vacuum pump. The addition of relatively inert gases, such as nitrogen or argon, stabilizes the plasma and controls the rate of ionization. Reactive oxygen species oxidize organic contaminants on the surface, creating volatile species that are pumped away. Etch rates are increased by providing more reactive species in the form of fluorine such as F_2, CF_4, or CHF_2.

Some laser-etched (and in some cases cleaned) parts are discussed next, with all kerf widths and hole diameters approximately 100 µm. Figure 9.2(a) shows a piece of polyimide about 125 µm thick that has been laser etched with a crosshair of about 100 µm kerf width. This part has been photographed directly from the laser and debris is clearly evident. Figure 9.2(b) shows the same part after gentle wiping with isopropyl alcohol (IPA). Note that the part is very clean.

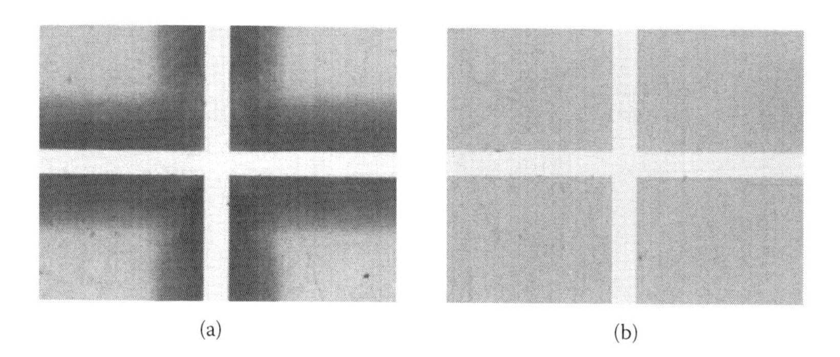

FIGURE 9.2
(a) Uncleaned and (b) cleaned laser-etched polyimide.

Figure 9.3(a) shows a piece of glass that has been laser etched and photographed with the debris being very evident. A gentle ultrasonic cleaning was performed and the resulting photograph is shown in Figure 9.3(b). The debris is almost entirely removed, but chip out of the brittle glass remains. It is sometimes possible to reduce this effect, but it is difficult to eliminate it without using a sacrificial layer or doing an additional postlaser grinding process.

Figure 9.4(a) shows a piece of stainless steel that has been laser etched and photographed (again, the debris is evident). Figure 9.4(b) shows the same part after cleaning with an IPA wipe, and Figure 9.4(c) shows the part after ultrasonic cleaning. It is evident that most of the debris can be removed; however, even after cleaning, this material shows another effect that may be undesirable: recast. The cleaned parts clearly show the melted/recast material around the cut and no gentle cleaning will remove this. In order to remove this recast effectively, a gentle electropolish or physical rub with an abrasive material is necessary.

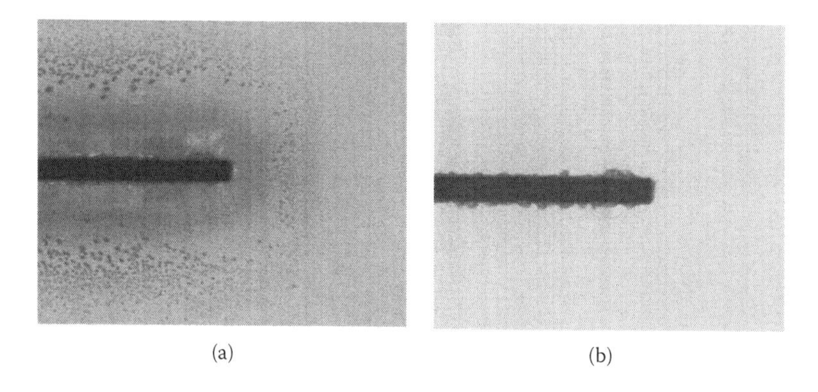

FIGURE 9.3
(a) Uncleaned and (b) cleaned laser-etched glass.

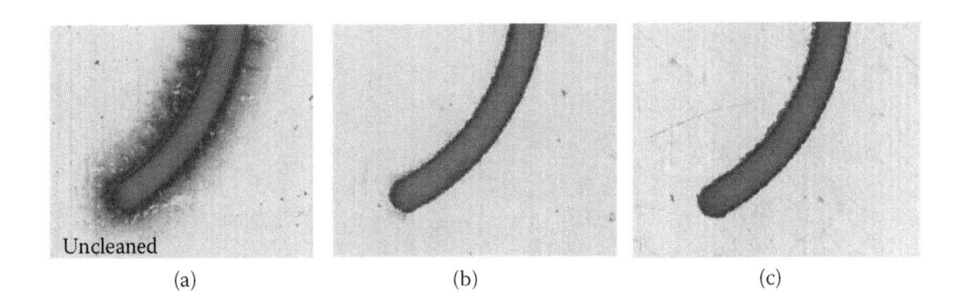

FIGURE 9.4
(a) Uncleaned stainless steel; (b) stainless steel cleaned with IPA wipe; (c) stainless steel cleaned ultrasonically.

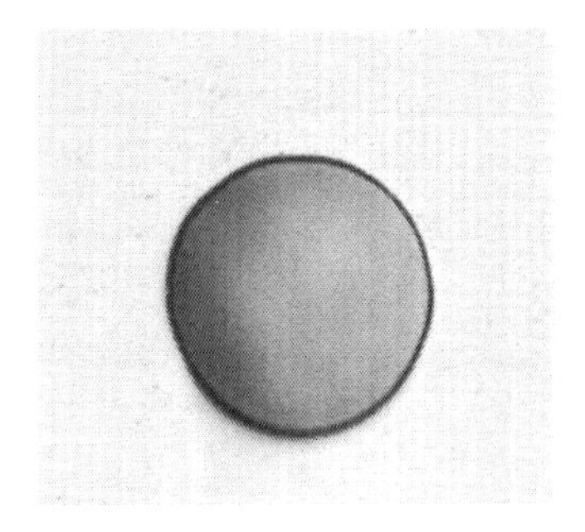

FIGURE 9.5
Electropolished stainless steel.

Figure 9.5 shows a stainless steel part that has been electropolished after laser drilling. No high spots are evident on this photograph; through the electropolishing process, the high spots have been preferentially removed.

9.2.5 Sacrificial Coatings

The simplest form of sacrificial coating is created by dipping the parts in soapy water, letting them dry, and then rinsing after laser processing. Another method is to spin photoresist over the whole surface and rinse after laser processing. Both methods will remove any loose debris generated during the laser process, but care must be taken to assure that the parts in question are not compromised by the coating or rinsing process.

Space does not permit an exhaustive review of all the postlaser cleaning techniques available, but the methods described before give an indication of some of the more common methods in use. Others, such as permanganate desmear, CO_2 "snow blasting," "lint tape" removal, etc., also exist and can be used to enhance the quality of laser-processed parts.

10

Conclusion

It has been said that physics is "the study of that which is interesting." If this is true, then the study of lasers must be "the study of that which is fascinating"! This text has tried to communicate at an understandable level the powerful utility of lasers in the micromachining of high-precision components for many different industries and in many different materials. A number of the topics discussed in this text could be further expanded to a full text in their own right, including many subjects that have only been touched upon, like welding, marking, surface modification, sensing, entertainment, and interferometry. Lasers are present or are an enabling technology in almost every aspect of modern life, but most of the time this is not even evident.

The text described or alluded to much of the history and development of high-precision micromachining lasers. Within the past decade, new laser types, such as fiber lasers, disk lasers, and ultrashort pulse lasers, have become available. More new laser types will be forthcoming. As part sizes and geometries become smaller, feature sizes will also need to be smaller and this means that more processing will be done with UV and short pulse lasers. Both near- and far-IR lasers will continue to be used heavily in many industries, but physics dictates that shorter wavelength photons will be needed in order to achieve smaller spot sizes on target. It is anticipated that frequency shifted fiber lasers will soon be commercially available, allowing attainment of these smaller feature sizes while taking advantage of the economic and engineering benefits of using fiber lasers (or some new laser source).

In keeping with the mantra, "smaller, faster, cheaper," new laser sources and optics will be available that have not even been currently envisioned. These lasers will be lightweight, cheaper (dollars per watt), and simpler and will have long MTBF (mean time between failures) while also having short MTR (mean time to repair). For micromachining, the dream laser is one that has a 100 fs pulse length, UV output (250 nm or so), high repetition rate (100 kHz or more), and high pulse energy (greater than 100 μJ/pulse and hopefully much more), and a cost of a few thousand dollars per watt of output power. This is the challenge to the laser manufacturers.

Appendix: Additional Reading

The following sources can be used to find more information and much more detail on certain subjects that, because of space limitations, cannot be fully covered in this book.

Boyd, R. W. 2003. *Non-linear optics,* 2nd ed. San Diego, CA: Academic Press, Inc.

Dickey, F. M. et al., eds. 2006. *Laser beam shaping applications.* Boca Raton, FL: Taylor & Francis Group.

Donnelly, J., and Massa, N. 2010. *Light—Introduction to optics and photonics.* Boston: New England Board of Higher Education.

Elliott, D. J. 1995. *Ultraviolet laser technology and applications.* San Diego, CA: Academic Press.

Laser Institute of America (www.lia.org). The LIA is a resource that can be used for information on lasers, safety, and applications. They sponsor many laser safety courses and conferences throughout the year and are accredited to instruct and certify laser safety officers.

Migliore, L. R. 1996. *Laser materials processing,* Vol. 46. New York: Marcel Decker Inc.

Ready, J., ed. 2001. *LIA handbook of laser materials processing.* Orlando, FL: Magnolia Publishing.

Siegman, A. E. 1986. *Lasers.* Mill Valley, CA: University Science Books.

Representative Publications

Angell, J., W. Ho, J. Bernstein, and R. Schaeffer. 1997. Processing parameters for laser micromachining. *Proceedings from Photonics West '97 Conference,* Society for Photo-optical Instrumentation Engineers, San Jose, CA, February 1997.

Angell, J., W. Ho, and R. Schaeffer. 1997. Effects of taper on drilling and cutting with a pulsed laser. *Photonics West 1997 Conference Proceedings,* SPIE, San Jose, CA, February 1997.

Angell, J., and R. Schaeffer. 1996. Laser processing of ceramics and CVD diamond film. *Proceedings from Advancements in the Application of Ceramics in Manufacturing,* Society of Manufacturing Engineers, Newton, MA, October 1996.

Bernstein, J., J. Guerette, and R. Schaeffer. 1997. Polyimide processing with a CW pumped q-switched frequency doubled Nd:YLF laser. *Proceedings from Photonics West '97 Conference,* Society for Photo-optical Instrumentation Engineers, San Jose, CA, February 1997.

Edwards, T., and R. Schaeffer. 2007. Micromachining with deep ultraviolet lasers: Cutting tool engineering. John Wm. Roberts & Associates, Northbrook, IL, October 2007.

Elmer, J. W., O. Yaglioglu, and R. D. Schaeffer. Direct patterning of vertically aligned carbon nanotube arrays using focused laser beam micromachining, to be published.

Fritz, D., S. Castaldi, F. Durso, R. Schaeffer, J. O'Connell, and G. Kardos. 1998. Limits of copper plating in high aspect ratio microvias. Presented at the Printed Circuits Conference, IPC, Long Beach, CA, April 1998, and published in CircuiTree, Campbell, CA, September 1998.

Hoult, T., and R. Schaeffer. 2002. Lasers—The best prescription for processing medical materials. *Medical Design News,* Penton Media, Cleveland, OH, April 2002.

Kardos, G., O. Derkach, and R. Schaeffer. 2006. 266 nm DPSS laser micromachining. *Industrial Laser Solutions,* Pennwell Publishing, Tulsa, OK, December 2006.

———. 2007. Precision laser drilling of alumina ceramic. *Industrial Laser Solutions,* Pennwell Publishing, Tulsa, OK, February 2007.

Kardos, G., and R. Schaeffer. 2006. Laser processing of flexible circuits. *PC Fab and Design,* UP Media, Inc., Marietta, GA, October 2006.

———. 2008. Comparison of fiber laser processing of some materials to UV processing. *Industrial Laser Solutions,* Pennwell Publishing, Tulsa, OK, February 2008.

———. 2008. Post laser cleaning techniques. *Industrial Laser Solutions,* Pennwell Publishing, Tulsa, OK, May 2008.

Lizotte, T., G. Kardos, and R. Schaeffer. 2008. Drilling of PCBs—An overview. *PC Fab and Design,* UP Media, Inc., Marietta, GA, June, 2008

Lovejoy, R. W., R. Schaeffer, D. L. Frasco, C. C. Chackerian, and R. W. Boese. 1985. Absolute line strengths of phosphine gas near 5 μm. *Journal of Molecular Spectroscopy* 109:246.

Ogura, G., R. Andrew, and R. Schaeffer. 1996. Practical consequences of matching real laser sources to target illumination requirements. *Proceedings from Photonics West '96 Conference* 2703:30–40. Society for Photo-optical Instrumentation Engineers, San Jose, CA, February 1996.

Ortabasi, U., D. Meier, J. Easoz, R. Schaeffer, M. Stepanova, W. Ho, J. Stokes, S. Drummer, J. Jafolla, and P. McKenna. 1997. Excimer micromachining for texturing silicon solar cells. *Photonics West 1997 Conference Proceedings,* SPIE, San Jose, CA, February 1997.

Schaeffer, R. 1995. Laser micromachining of disposable medical devices. *Proceedings from Manufacturing Medical Plastics '95 Conference,* Society of Manufacturing Engineers and the Plastics and Molders and Manufacturers Group, Chicago, IL August 1995.

———. 1995. Novel high-power Nd:YLF laser for CVD-diamond micromachining. *Proceedings from Micromachining and Microfabrication Process Technology* 2639:325–334, Society for Photo-optical Instrumentation Engineers, Austin, TX.

———. 1996. Quality control issues in laser micromachining. *Proceedings from MD&M West '96 Conference,* Cannon Communications, Anaheim, CA, February 1996.

———. 1996. Laser-based dielectric material removal. *Industrial Laser Review,* Pennwell Publishing, Tulsa, OK, December 1996.

———. 1996. Laser-manufactured features in medical catheters and angioplasty devices. Medical Device and Diagnostics Industry, November 1996.

———. 1996. Laser micromachining of medical devices. *Medical Plastics and Biomaterials* May/June 1996.

———. 1997. Excimer lasers…unique manufacturing tools. *Job Shop Technology* January/March 1997.

———. 1997. Laser micromachining: High speed hole drilling for electronics packaging (in Japanese–English translations available). *Japanese Electronics Technology* April 1997.
———. 1998. An overview of laser microvia drilling. *Future Circuits International,* Issue 3, Technology Publishing, Ltd., London, England.
———. 1998. A status report on laser micromachining. *Industrial Laser Review* 13 (9), Pennwell Publishing, Tulsa, OK, September 1998.
———. 1998. *Laser microvia drilling—Recent advances.* Campbell, CA: CircuiTree Publishing.
———. 1998. Marking medical products with lasers. *Proceedings of MD&M East,* New York, Canon Communications, June 1998.
———. 1999. A radiant solution—Using laser micromachining to drill high quality holes in ceramic substrates. *Ceramic Industry,* Business News Publishing Corp., Northbrook, IL, June 1999.
———. 1999. The case for laser microvia drilling. *Industrial Laser Solutions,* Pennwell Publishing, Tulsa, OK, March, 1999.
———. 2000. Basic laser physics (Lasers 101). Published in the column "Seeing the Light," CircuiTree, Business News Publishing Company, Northbrook, IL, September, 2000.
———. 2000. Chapters published in *LIA handbook of laser materials processing,* ed. John F. Ready and Dave F. Farson. Orlando, FL: Laser Institute of America and Magnolia Publishing, Inc.
———. 2000. How to decide on commercially available microvia drillers. Published in the column "Seeing the Light," CircuiTree, Business News Publishing Company, Northbrook, IL, December, 2000.
———. 2000. Material/photon interaction. Published in the column "Seeing the Light," CircuiTree, Business News Publishing Company, Northbrook, IL, October, 2000.
———. 2000. Micromachining technology. PhotoMachining, Inc. Micromachining seminar, course notes updated 2000.
———. 2000. Preparing your facility for laser tools. Published in the column "Seeing the Light," CircuiTree, Business News Publishing Company, Northbrook, IL, November, 2000.
———. 2001. A closer look at laser ablation, *Industrial Laser Solutions,* Pennwell Publishing, Tulsa, OK, September, 2000 and re-published in *Laser Focus World,* Pennwell Publishing, Tulsa, OK, June 2001.
———. 2001. A visit with CEMCO, published in February CircuiTree Web site, CircuiTree, Business News Publishing Company, Northbrook, IL, February, 2001.
———. 2001. Lasers for board testing. Published in the column "Seeing the Light," CircuiTree, Business News Publishing Company, Northbrook, IL, January, 2001.
———. 2001. Lasers for direct imaging. Published in the column "Seeing the Light," CircuiTree, Business News Publishing Company, Northbrook, IL, February, 2001.
———. 2001. Lasers in related microelectronic related fields: Part I. Published in the column "Seeing the Light," CircuiTree, Business News Publishing Company, Northbrook, IL, April, 2001.
———. 2001. Lasers in related microelectronic related fields: Part II. Published in the column "Seeing the Light," CircuiTree, Business News Publishing Company, Northbrook, IL, May, 2001.

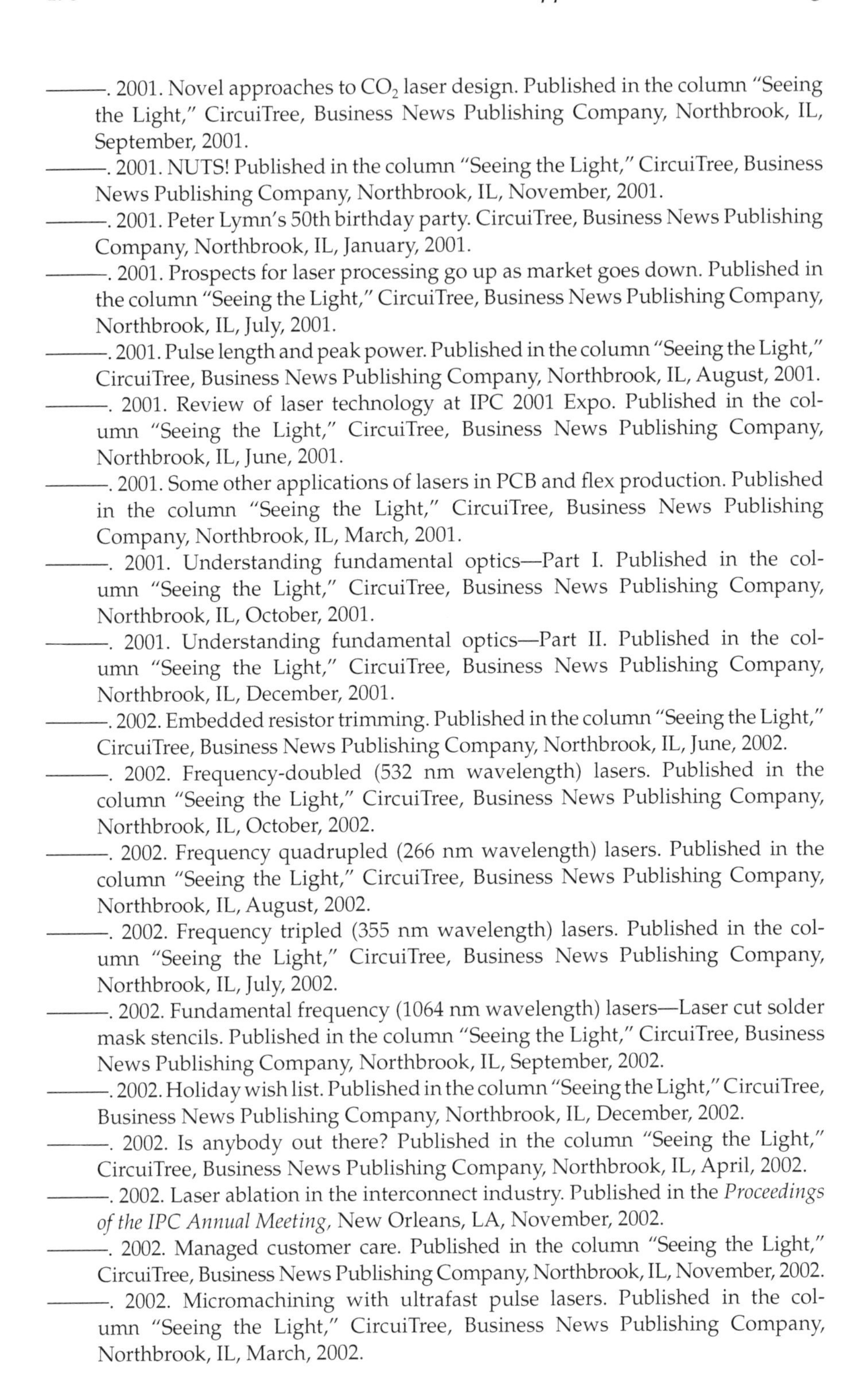

———. 2001. Novel approaches to CO_2 laser design. Published in the column "Seeing the Light," CircuiTree, Business News Publishing Company, Northbrook, IL, September, 2001.
———. 2001. NUTS! Published in the column "Seeing the Light," CircuiTree, Business News Publishing Company, Northbrook, IL, November, 2001.
———. 2001. Peter Lymn's 50th birthday party. CircuiTree, Business News Publishing Company, Northbrook, IL, January, 2001.
———. 2001. Prospects for laser processing go up as market goes down. Published in the column "Seeing the Light," CircuiTree, Business News Publishing Company, Northbrook, IL, July, 2001.
———. 2001. Pulse length and peak power. Published in the column "Seeing the Light," CircuiTree, Business News Publishing Company, Northbrook, IL, August, 2001.
———. 2001. Review of laser technology at IPC 2001 Expo. Published in the column "Seeing the Light," CircuiTree, Business News Publishing Company, Northbrook, IL, June, 2001.
———. 2001. Some other applications of lasers in PCB and flex production. Published in the column "Seeing the Light," CircuiTree, Business News Publishing Company, Northbrook, IL, March, 2001.
———. 2001. Understanding fundamental optics—Part I. Published in the column "Seeing the Light," CircuiTree, Business News Publishing Company, Northbrook, IL, October, 2001.
———. 2001. Understanding fundamental optics—Part II. Published in the column "Seeing the Light," CircuiTree, Business News Publishing Company, Northbrook, IL, December, 2001.
———. 2002. Embedded resistor trimming. Published in the column "Seeing the Light," CircuiTree, Business News Publishing Company, Northbrook, IL, June, 2002.
———. 2002. Frequency-doubled (532 nm wavelength) lasers. Published in the column "Seeing the Light," CircuiTree, Business News Publishing Company, Northbrook, IL, October, 2002.
———. 2002. Frequency quadrupled (266 nm wavelength) lasers. Published in the column "Seeing the Light," CircuiTree, Business News Publishing Company, Northbrook, IL, August, 2002.
———. 2002. Frequency tripled (355 nm wavelength) lasers. Published in the column "Seeing the Light," CircuiTree, Business News Publishing Company, Northbrook, IL, July, 2002.
———. 2002. Fundamental frequency (1064 nm wavelength) lasers—Laser cut solder mask stencils. Published in the column "Seeing the Light," CircuiTree, Business News Publishing Company, Northbrook, IL, September, 2002.
———. 2002. Holiday wish list. Published in the column "Seeing the Light," CircuiTree, Business News Publishing Company, Northbrook, IL, December, 2002.
———. 2002. Is anybody out there? Published in the column "Seeing the Light," CircuiTree, Business News Publishing Company, Northbrook, IL, April, 2002.
———. 2002. Laser ablation in the interconnect industry. Published in the *Proceedings of the IPC Annual Meeting,* New Orleans, LA, November, 2002.
———. 2002. Managed customer care. Published in the column "Seeing the Light," CircuiTree, Business News Publishing Company, Northbrook, IL, November, 2002.
———. 2002. Micromachining with ultrafast pulse lasers. Published in the column "Seeing the Light," CircuiTree, Business News Publishing Company, Northbrook, IL, March, 2002.

———. 2002. Strip that solder mask. Save those boards! Published in the column "Seeing the Light," CircuiTree, Business News Publishing Company, Northbrook, IL, May, 2002.
———. 2002. Understanding fundamental optics—Part III. Published in the column "Seeing the Light," CircuiTree, Business News Publishing Company, Northbrook, IL, January, 2002.
———. 2002. When is the right time to buy a laser tool? Published in the column "Seeing the Light," CircuiTree, Business News Publishing Company, Northbrook, IL, February, 2002.
———. 2003. A very short essay on erbium-doped fiber amplifiers. Published in the column "Seeing the Light," CircuiTree, Business News Publishing Company, Northbrook, IL, May, 2003.
———. 2003. Diode lasers for communications. Published in the column "Seeing the Light," CircuiTree, Business News Publishing Company, Northbrook, IL, March, 2003.
———. 2003. Experience required. Published in the column "Seeing the Light," CircuiTree, Business News Publishing Company, Northbrook, IL, December 2003.
———. 2003. Fiber Bragg gratings. Published in the column "Seeing the Light," CircuiTree, Business News Publishing Company, Northbrook, IL, February, 2003.
———. 2003. Foolish things I have or have not done. Published in the column "Seeing the Light," CircuiTree, Business News Publishing Company, Northbrook, IL, April, 2003.
———. 2003. Galvo-based laser scanning systems. Published in the column "Seeing the Light," CircuiTree, Business News Publishing Company, Northbrook, IL, September, 2003.
———. 2003. Purchasing laser services. Published in the column "Seeing the Light," CircuiTree, Business News Publishing Company, Northbrook, IL, August, 2003.
———. 2003. Sick minds. Printed circuit design and manufacture. GMP Media, Inc., Marietta, GA, May, 2003.
———. 2003. So, you want to know more about lasers? Published in the column "Seeing the Light," CircuiTree, Business News Publishing Company, Northbrook, IL, January, 2003.
———. 2003. Thank God for the medical market. Published in the column "Seeing the Light," CircuiTree, Business News Publishing Company, Northbrook, IL, December, 2003.
———. 2003. The collapse of the old man in the mountain. Published in the column "Seeing the Light," CircuiTree, Business News Publishing Company, Northbrook, IL, July, 2003.
———. 2003. Travel jitters. Published in the column "Seeing the Light," CircuiTree, Business News Publishing Company, Northbrook, IL, June, 2003.
———. 2004. Current trends in laser microvia drilling. *PC Fab and Design,* UP Media, Inc., Marietta, GA, April, 2004.
———. 2004. Diffractive beam shaping improves laser material processing applications. Published in the column "Seeing the Light," CircuiTree, Business News Publishing Company, Northbrook, IL, December, 2004.
———. 2004. Formulas, formulas. Published in the column "Seeing the Light," CircuiTree, Business News Publishing Company, Northbrook, IL, March, 2004.

———. 2004. It's Expo time again! Published in the column "Seeing the Light," CircuiTree, Business News Publishing Company, Northbrook, IL, February, 2004.
———. 2004. New BGA solder mask repair technique using laser-cut stencils. Published in the column "Seeing the Light," CircuiTree, Business News Publishing Company, Northbrook, IL, August, 2004.
———. 2005. A good economic indicator. Published in the column "Seeing the Light," CircuiTree, Business News Publishing Company, Northbrook, IL, June, 2005.
———. 2005. Fiber lasers: The hot new laser source. Published in the column "Seeing the Light," CircuiTree, Business News Publishing Company, Northbrook, IL, March, 2005.
———. 2005. Review: Old markets and new. Published in the column "Seeing the Light," CircuiTree, Business News Publishing Company, Northbrook, IL, December, 2005.
———. 2005. SuperPulse laser. Published in the column "Seeing the Light," CircuiTree, Business News Publishing Company, Northbrook, IL, September, 2005.
———. 2006. They didn't teach this in business school. *Industrial Laser Solutions,* Pennwell Publishing, Tulsa, OK, Two-part feature article in June/July, 2006.
———. 2007. Another one bites the dust! Published in the I-Connect007 (PCB007) web column "Light Reading," I-Connect007 (http://www.pcb007.com/RonSchaeffer.aspx?z=192), Seaside, OR, July, 2007.
———. 2007. I'm back! Published in the I-Connect007 (PCB007) web column "Light Reading," I-Connect007 (http://www.pcb007.com/RonSchaeffer.aspx?z=192), Seaside, OR, June, 2007.
———. 2007. Safety in the laser lab part I—Practical considerations! Published in the I-Connect007 (PCB007) web column "Light Reading," I-Connect007 (http://www.pcb007.com/RonSchaeffer.aspx?z=192), Seaside, OR, September, 2007.
———. 2007. Thoughts on attending conferences! Published in the I-Connect007 (PCB007) web column "Light Reading," I-Connect007 (http://www.pcb007.com/RonSchaeffer.aspx?z=192), Seaside, OR, August, 2007.
———. 2008. Post laser cleaning techniques. Published in the I-Connect007 (PCB007) web column "Light Reading," I-Connect007 (http://www.pcb007.com/RonSchaeffer.aspx?z=192), Seaside, OR, Seaside, OR, June, 2008.
———. 2008. Safety in the laser lab part II—Standards. Published in the I-Connect007 (PCB007) web column "Light Reading," I-Connect007 (http://www.pcb007.com/RonSchaeffer.aspx?z=192), Seaside, OR, May, 2008.
———. 2008. Semicon 2008. Published in the I-Connect007 (PCB007) web column "Light Reading," I-Connect007 (http://www.pcb007.com/RonSchaeffer.aspx?z=192), Seaside, OR, July, 2008.
———. 2009. Cost considerations when buying a laser. Published in the column "LASERpoints" in *MicroManufacturing Magazine,* jwr, Northbrook, IL, Winter, 2009.
———. 2009. Laser machining: Basics and benefits. Published in the column "LASERpoints" in *MicroManufacturing Magazine,* jwr, Northbrook, IL, Spring, 2009.
———. 2009. MicroMachining with lasers. *Photonics Tech Briefs,* Tech Briefs Media Group, January, 2009.
———. 2009. Minimizing HAZ when laser micromachining. Published in the column "LASERpoints" in *MicroManufacturing Magazine,* jwr, Northbrook, IL, Summer, 2009.

———. 2009. Outsourcing laser work: Putting the cost puzzle together. Published in the column "LASERpoints" in *MicroManufacturing Magazine,* jwr, Northbrook, IL, Fall, 2009.

———. 2010. Basics of lasering high aspect ratio holes. Published in the column "LASERpoints" in *MicroManufacturing Magazine,* jwr, Northbrook, IL, September/October, 2010.

———. 2010. Fiducials: How to be where you need to be. Published in the column "LASERpoints" in *MicroManufacturing Magazine,* jwr, Northbrook, IL, March/April, 2010.

———. 2010. Lasers contribute to next-gen microfluidic devices. Published in the column "LASERpoints" in *MicroManufacturing Magazine,* jwr, Northbrook, IL, November/December, 2010.

———. 2010. Lasers make their mark on microparts. Published in the column "LASERpoints" in *MicroManufacturing Magazine,* jwr, Northbrook, IL, Jan/Feb, 2010.

———. 2010. The long and ultrashort of laser pulses. Published in the column "LASERpoints" in *MicroManufacturing Magazine,* jwr, Northbrook, IL, July/August, 2010.

———. 2010. Understanding and controlling taper when laser machining. Published in the column "LASERpoints" in *MicroManufacturing Magazine,* jwr, Northbrook, IL, May/June, 2010.

———. 2011. Gas gives big assist to many lasing jobs. Published in the column "LASERpoints" in *MicroManufacturing Magazine,* jwr, Northbrook, IL, January/February 2011.

———. 2011. Ultra short pulse laser processing. *Medical Design Technology* Advantage Media, Rockaway, NJ, March 2011.

———. 2011. Laser patterning of thin films. Published in the column "LASERpoints" in *MicroManufacturing Magazine,* jwr, Northbrook, IL, March/April, 2011.

———. 2011. Laser drilling really small holes. Published in the column "LASERpoints" in *MicroManufacturing Magazine,* jwr, Northbrook, IL, May/June, 2011.

———. 2011. Laser micromachining of polymers. *Materials World,* published by the Institute of Materials, Minerals and Mining, London, UK, August, 2011.

———. 2011. Post laser cleaning methods. Published in the column "LASERpoints" in *MicroManufacturing Magazine,* jwr, Northbrook, IL, September/October, 2011.

———. 2011. Multiple laser beams from one galvo. *Industrial Laser Solutions,* Pennwell Publishing, Tulsa, OK, November/December, 2011.

———. 2011. High resolution, low taper excimer laser machining of thick materials, NASA SBIR Contract #NAS5-38042, Phase I Final Report.

Schaeffer, R., and J. Angell. 1997. A promise of cost effective solutions in microelectronics manufacturing. *Photonics West 1997 Conference Proceedings,* SPIE, San Jose, CA, February 1997.

Schaeffer, R., L. Chen, and W. Ho. 1996. Laser planarization of chemical vapor deposited diamond film. *Proceedings from Photonics West '96 Conference,* Society for Photo-optical Instrumentation Engineers, San Jose, CA, February 1996.

Schaeffer, R., and D. Grossman. 2003. Medical device manufacturing. *Industrial Laser Solutions,* Pennwell Publishing, Tulsa, OK, November, 2003.

Schaeffer, R., and T. Hannon. 2001. Micromachining in the UV. *Laser Focus World,* Pennwell Publishing Company, Tulsa, OK, February, 2001.

Schaeffer, R., and G. Kardos. 2000. *Laser repair of printed circuit boards*. CircuiTree, Business News Publishing Company, Northbrook, IL, August, 2000.

Schaeffer, R., G. Kardos, and O. Derkach. 2002. Laser processing of glass. *Industrial Laser Solutions*, Pennwell Publishing, Tulsa, OK, September, 2002.

———. 2004. Laser micromachining of thin metals. Published in the *Proceedings of ICALEO 2004*, Laser Institute of America.

———. 2004. Laser micromachining of thin metals—Cutting speed and quality comparison. *Industrial Laser Solutions*, Pennwell Publishing, Tulsa, OK, December, 2004.

Schaeffer, R., G. Kardos, O. Derkach, C. Dunsky, L. Migliore, and A. Orkan. 2004. Laser choices for micromachining glass and other hard, brittle materials. Published in the *Proceedings of ICALEO 2004*, Laser Institute of America.

Schaeffer, R., G. Kardos, and J. Keating. 2002. Outsourcing of laser drilling. *PC Fab*, GMP Media, Inc., Marietta, GA, June, 2002.

Schaeffer, R., G. Kardos, S. Murphy, and D. Grossman. 2008. Some observations of effects using short pulse lasers for micromachining. *Industrial Laser Solutions*, Pennwell Publishing, Tulsa, OK, November, 2008.

Schaeffer, R., and R. W. Lovejoy. 1985. Absolute line strengths of $^{74}GeH_4$ near 5 μm. *Journal of Molecular Spectroscopy* 113:310.

Schaeffer, R., R. W. Lovejoy, W. B. Olson, and G. Tarrago. 1988. Analysis of the high resolution spectrum of $^{28}SiH_3D$ from 1450 to 1710 cm^{-1}. *Journal of Molecular Spectroscopy* 128:135.

Schaeffer, R., T. P. McGarry, and M. J. Scaggs. 1990. Materials processing with excimer lasers. *Materials and Manufacturing Processes* 5 (4).

Schaeffer, R., and J. O'Connell. 2002. Comparison of some lasers used for micromachining plastics. *Industrial Laser Solutions*, Pennwell Publishing, Tulsa, OK, May, 2002.

Schaeffer, R., J. O'Connell, M. Gitin, E. Rea, A. Caprara, and G. Nazary. 1997. The use of solid state lasers to replace traditional UV photon sources in medical device manufacturing. *ICALEO Conference Proceedings*, Laser Institute of America, San Diego, CA, November, 1997.

Schaeffer, R., and T. Pflanz. 2008. Micromachining with UV lasers—Shorter pulses and/or shorter wavelength? *Medical Device & Diagnostics Industry*, Canon Publishing, Los Angeles, CA, August 2008.

Schaeffer, R., and D. Schaefer. 1989. ISLE—Selective laser etching. Published in the column "LASERpoints" in *MicroManufacturing Magazine*, jwr, Northbrook, IL, July/August, 2011.

Schaeffer, R., J. Sproul, J. O'Connell, C. Van Vloten, and A. W. Mantz. 1989. Multipass absorption cell for low concentration H_2O analysis using a Pb-salt: Tunable diode laser spectrometer. *Applied Optics* 28:1710.

Schaeffer, R., and I. Syrgabaev. 2001. Lasers used in the production of solar panels. *Industrial Laser Solutions*, Pennwell Publishing, Tulsa, OK, January, 2001.

Wall, D., and R. Schaeffer. 1997. Using lasers for leak test validation in medical device manufacturing. *MDM East Conference Proceedings*. Canon Communications, Anaheim, CA, June 1997.

Problems

These questions and answers are intended to further the understanding of the topics presented in the text. Some answers are 'exact' while others depend on assumptions and are estimates. The thought process behind all of the answers is fully documented, but there may be, in some cases, equally valid solutions to problems. The author is always interested in any thoughts, corrections, comments, or ideas that may come to mind.

The answers appear in the book's solution manual.

CHAPTER 1 - INTRODUCTION

Q1) I want to drill a 10 micron diameter hole in a 20 micron thick piece of material. Which laser would NOT be appropriate?

a) 248 nm excimer
b) 355 nm DPSS
c) 266 nm DPSS
d) CO_2

Q2) A piece of alumina ceramic 200 microns thick is to be cut. The kerf width is 200 microns. Which method would NOT be appropriate?

a) CO_2 laser cutting
b) 355 nm DPSS
c) Wire EDM
d) Mechanical Saw

CHAPTER 2 – LASER THEORY AND OPERATION

Q1) A CO_2 laser has a wavelength of about 10 microns, in the IR portion of the spectrum. What is the associated photon energy (E_{photon})?

a) 2.2×10^{-21} J
b) 3.5×10^{-35} J/s
c) 5.6×10^{10} J
d) None of the above.

Q2) What is the frequency of a 355 nm photon?

a) 5.54×10^{-10} s^{-1}

b) 8.45×10^{14} s-1

c) 5.54×10^{10} s

d) 8.45×10^{14} s

Q3) A Four level laser transition has the upper lasing level at an energy level of 30.1×10^{-20} J above the ground state and the lower level of the lasing transition at an energy of 11.4×10^{-20} J. What is the wavelength of the photon generated from this transition?

a) 1.06 microns

b) 10 microns

c) 355 nm

d) 266 nm

Q4) A certain gas is contained in a cylinder with windows on each end. The concentration of this gas is fixed. A beam of light is passed through the gas and it is found that the laser light is attenuated by 90%. Some gas is let out of the sample chamber and the light is again passed through – this time showing an attenuation of only 40%. What is the concentration of gas in the second experiment?

a) 22.5 c_1

b) 3.3×10^{23} cm^{-3}

c) 0.444 c_1

d) There is no gas left in the container

Q5) Methane is a molecule containing one carbon atom and 4 hydrogen atoms arranged in a tetrahedral configuration. How many vibrational modes does this molecule contain?

a) 1

b) 5

c) 9

d) 15

Q6) Which of the following applications would NOT be suitable for a UV laser?

a) welding two dissimilar metals
b) drilling very small (less than 25 microns diameter) holes
c) cutting thin plastics
d) marking anodized Al

Q7) My laser has a pulse energy of 300 mJ and a pulse length of 20 ns. What is the Peak Power of this laser?

a) 50 W
b) 1500 W
c) 15 MW
d) 1.5 mW

Q8) Assuming I image all of the laser energy in the above example into a round spot of diameter 10 microns, what is the Peak Power Intensity?

a) 1.91×10^{13} J/s-cm^2
b) 1.91×10^{13} W/cm2
c) Both of the above are correct
d) None of the above are correct

Q9) In Figure 2.8, what is the approximate pulse length of this laser?

a) 5 ps
b) 5 ns
c) 50 fs
d) 50 ns

Q10) Match each of the lasers below with the proper corresponding statement.

a) CO_2 laser	w) Small features, good edge quality
b) Fundamental fiber laser	x) Some melt on edges
c) Excimer laser	y) Transparent material, clean cuts
d) 355 nm picosecond laser	z) Not useful on most polymers

Q11) Which statement about the CO_2 laser is incorrect?

a) For low power applications it is the cheapest laser to buy ($/W).
b) Their large size precludes their use in most clean room environments.
c) The gas mix contains helium and nitrogen.
d) The first order material interaction is via a thermal process.

Q12) Which statement about the excimer laser is incorrect?

a) Excimer lasers use and generate toxic gases that must be dealt with.
b) Excimer lasers are relatively large in size.
c) Excimer lasers are best utilized in a focal point machining configuration.
d) Excimer lasers can be used to generate very high resolution features.

Q13) Which statement about the fiber laser is correct?

a) Fiber lasers are only available in power output up to about 100 W.
b) Fiber lasers are generally more costly than other lasers to operate.
c) Fiber lasers have output wavelengths in the near IR.
d) Fiber lasers have wall plug efficiencies less than 5%.

Q14) Which statement about USP lasers is correct?

a) USP lasers are used in applications requiring the best edge quality.
b) USP lasers are relatively inexpensive on a $/W basis.
c) USP lasers are only available with fundamental wavelength output.
d) USP lasers do not require cooling because of their high wall plug efficiency.

CHAPTER 3 - OPTICS

Q1) A scuba diver swimming underwater shines her flashlight out of the water. If the light strikes the surface of the water (n = 1.33) at an angle of 38 degrees from normal, what will the angle of refraction be?

a) 38 degrees
b) 360 degrees
c) 55 radians

d) 55 degrees

Q2) What is the speed of light (m/s) in the cornea of the eye (n = 1.37)?

a) 4.11×10^8 m/s
b) 2.2×10^8 m/s
c) 3.0×10^8 m/s
d) No change – light always travels at the same speed.

Q3) The same scuba diver as in question 1 above shines the same light at the same angle, only in this case there is a reflective oil film on the surface of the water. What happens to the light?

a) It exits the water and is refracted at an angle of 55 degrees from the normal.
b) It exits the water and is refracted at an angle of 38 degrees from the normal
c) It is reflected back into the water at an angle of 55 degrees from the normal.
d) It is reflected back into the water at an angle of 38 degrees from the normal.

Q4) Figure 3.2 shows a positive spherical lens. This lens can be used to focus collimated incoming light or to image, as shown. Assuming the light is coming from infinitely far away, how far away from the lens will the beam focus?

a) The light will not focus – it will remain collimated into infinity on the exit side.
b) The light will focus at the focal point of the lens.
c) The light will focus at the image plane of the lens.
d) The light will focus at the object plane.

Q5) Figure 3.3 shows a negative spherical lens. Assuming the light is coming from infinity and passes through this lens, where will the light focus?

a) The light will not focus, it will disperse.
b) The light will have a virtual focus on the incoming side of the lens.
c) The dispersion angel will depend on the virtual focus of the lens.
d) All of the above are correct.

Q6) Which of the following statements about the positive cylinder lens shown in Figure 3.4 are untrue?

a) The resulting image of a square will be a square.
b) The resulting image of a square will be a rectangle.
c) The energy density in the resulting image will be greater than in the incoming beam.
d) At the focal point of the lens, the light will be oriented in a line.

Q7) Prisms are very useful for all of the following except one.

a) Breaking up light into its constituent components.
b) Beam steering off reflective surfaces.
c) Changing the polarization of the incoming light.
d) Beam steering from refraction.

Q8) The magnifying power of a telescope is found by taking the ratio of the focal length of the two lenses regardless of whether the design is Keplerian or Galilean. In each case, if the ratio is less than 1, the image is smaller and if it is more than 1, the image is larger (depending on beam incoming direction). I have a 300 mm positive spherical lens and I want to use it to make a 3× telescope. What is the focal length of the second lens?

a) 300 mm
b) 150 mm
c) 100 mm
d) 600 mm

Q9) In question #8 above, what is the distance between the elements if a positive lens is used for the second lens? What is the distance between the elements if a negative lens is used for the second lens?

a) 400 mm and 100 mm
b) 200 mm and 300 mm
c) 400 mm and 200 mm
d) 100 mm and 400 mm

Q10) Which of the statements below is correct?

a) When using a telescope, if the optical elements are placed at exactly the sum of the focal distances of both lenses, the beam will be slightly divergent.
b) When using a telescope, if the optical elements are placed at exactly the sum of the focal distances of both lenses, the beam will be slightly convergent.
c) Telescopes cannot in general be used to both expand AND contract the beam.
d) AR coatings help improve throughput efficiency on telescope optics.

Q11) Homogenizers are best used when:

a) The optical set up is fixed and not subject to changes during operation.
b) Beam uniformity is not a critical factor.
c) There is a limited budget.
d) Both a and c are correct.

Q12) In materials processing applications, laser light that is not circularly polarized can result in the following:

a) A significant reduction in the absorption of any material at a given wavelength.
b) Different line widths when cutting in different directions.
c) Potential destruction of down stream optics.
d) Both a and b are correct.

Q13) What is the smallest attainable spot size on target using a CO_2 laser and a one inch diameter, 200 mm focal length lens in a fixed beam delivery system? Assume the M^2 is given by the manufacturer of the laser as 1.4.

a) 140 microns
b) 200 microns
c) 100 microns
d) 70 microns

Q14) In the example above, what is the spot size using a galvanometer beam delivery with an f-theta lens of focal length 300 mm and a 30 mm aperture ?

a) 314 microns
b) 266 microns
c) 196 microns
d) 156 microns

Q15) A field size of 30 inches is required on target for a 3D galvanometer based beam delivery system. The tool uses a 355 nm laser with M^2 of 1.2. The input aperture of the galvo is 14 mm and the measured spot size on target is 100 microns. How far above the work surface is the lens?

a) 1730 microns
b) 1370 microns
c) 1370 mm
d) 1730 mm

Q16) An imaging system uses a 100 mm imaging lens. The mask is 300 mm away from the lens. How far is the image from the lens?

a) 300 mm
b) 105 mm
c) 150 mm
d) 100 mm

Q17) In the above example, what is the demag?

a) 4
b) 2
c) 3
d) 1

Q18) An incoming excimer laser beam has about 300 mJ total energy in a rectangular beam of size approximately 2 cm × 1 cm. A mask is placed in the beam 300 mm in front of a spherical imaging lens. The total demagnification of the beam delivery system is 5×. Assuming an ideal condition of no optic losses, what is the energy density on target?

a) 3750 mJ/cm^2
b) 37.5 mJ/mm^2
c) 3.75 J/cm^2
d) All of the above are correct!

Q19) The same laser as above is used but this time with a cylindrical lens that compresses the beam in the short direction. What is the resulting beam size on target?

a) 4 mm × 10 mm
b) 4 mm × 2 mm
c) 20 mm × 2 mm
d) 20 mm × 4 mm

Q20) In the example above, what is the resulting energy density on target (ignoring optics losses)?

a) 300 mJ/cm^2
b) 3750 mJ/cm^2
c) 150 mJ/cm^2
d) 750 mJ/cm^2

Q21) I have a non-circular laser beam with a usable, reasonably homogeneous area of about 5 mm × 5 mm square. I want to image a rectangle on target using a mask that is about 2 mm × 4 mm. What is the beam utilization on this optical set up?

a) 32%
b) 32 mJ
c) 32 microns
d) 32 mm^2

Q22) In the above example, what would be an easy way to increase the beam utilization using only two additional optics?

a) Use a spherical 2× telescope before the mask
b) Use a spherical 2× telescope after the mask

c) Use a cylindrical 2× telescope before the mask
d) Use a cylindrical 2× telescope after the mask

Q23) A coordinated opposing motion system allows a large mask to be scanned thereby generating a large area on target with high resolution features. In a 5× demag beam delivery system, how fast does the lower stage move if the mask stage moves at 100 mm/sec?

a) 100 mm/sec
b) 20 mm/sec
c) 500 mm/sec
d) 50 mm

Q24) Assuming we use the Usable Fraction of an Excimer beam (in other words, assume the beam is homogeneous), and we want to split the beam into 6 parts using physical 'scraping' optics, what is the resulting energy density in each of the beams. Assume the UFB is 1 cm × 2 cm and the total laser energy per pulse is 500 mJ. Assume further that the optical demag of the system is 10×.

a) The example is over specified and cannot be solved.
b) It depends on the wavelength of the laser.
c) 50 J/cm^2
d) 250 mJ/cm^2

CHAPTER 4 – LIGHT – MATERIAL INTERACTION

Q1) In Figure 4.1, energy from photons that hit the target at an energy density below the ablation threshold can do all of the following except?

a) Impart heat into the material.
b) Reflect from the surface.
c) Mark the surface.
d) Create an acoustic shock as they hit the surface.

Q2) Refer to Figure 4.2. Assuming a UV laser uniformly removes 0.1 micron of material per pulse, how many pulses will it take to percussion drill a hole in a 100 micron thick piece of material?

a) 1,000
b) 100
c) 10
d) 1

Q3) Which of the factors below can be used to minimize HAZ?

a) Work at a higher repetition rate of the laser.
b) Use a longer pulse length laser.
c) Use a homogeneous beam rather than Gaussian.
d) Find a material with lower absorption.

Q4) In Figure 4.2 the calculated spot size is shown as d. However, in practice the actual spot size on target will be somewhat different depending on the absorption characteristics of the material being processed. What factors might cause the actual spot on target to be SMALLER than the calculated value?

a) A mistake was made in the calculation.
b) The material has a high ablation threshold.
c) Both of the above.
d) None of the above.

Q5) Refer to Figure 4.3. Assume that the taper angle is 5 degrees on each side and that the material is 100 microns thick. If I want a 50 micron exit hole diameter, what will be the entrance hole diameter?

a) 32.5 microns
b) 67.5 microns
c) 41.25 microns
d) None of the above.

Q6) Which of the below would probably not be possible using a laser?

a) Drill a 10 micron diameter hole in 1mm thick stainless steel.
b) Drill a 10 micron diameter hole in 100 micron thick polyimide.

c) Drill a 100 micron diameter hole in 100 micron thick stainless steel.
d) They are all possible and in fact quite easily doable.

CHAPTER 5 – SYSTEM INTEGRATION

Q1) Which of the following statements is true?

a) The correct choice of a laser source assures good processing results on target.
b) Lasers are not complementary to other existing technologies.
c) Lasers are best utilized in non- manufacturing environments.
d) Lasers can sometimes be the only viable solution to an existing manufacturing problem.

Q2) What elements below are part of a laser beam delivery system?

a) Beam enclosure.
b) Focusing and imaging optics.
c) Turning optics.
d) All of the above.

Q3) Which of the following statements are untrue? Tooling is very important in materials processing because:

a) Small parts must be held firmly and without distortion if high tolerances are to be met.
b) Proper tooling can speed the total accumulated cycle time.
c) Tooling is disposable and therefore cheap.
d) Tooling helps to minimize operator error.

Q4) Which statement concerning assist gases is true?

a) Nitrogen is a good assist gas because it has the potential to store a lot of excess energy.
b) Argon makes a good cover gas because it is a small, light atom.
c) Helium makes a good assist gas because it is a large atom and can store energy effectively.
d) Introduction of oxygen into the assist gas will tend to quench burning.

Q5) Safety goggles should be worn at ALL times when using a:

a) Class IV laser
b) Class I laser
c) Mechanical drill
d) Both a and c

Q6) What is NOT a good way to avoid exposure from stray laser light?

a) Always work with a Class I tool.
b) Wear long sleeve clothing and no jewelry.
c) Close your eyes and cross your fingers.
d) Apply sunscreen to any exposed skin areas.

Q7) Why is a 532 nm laser very dangerous?

a) The UV wavelength is harmful to skin and eyes.
b) The invisible beam makes it hard to use, especially when aligning and performing maintenance.
c) These lasers require extremely large power supplies and can create strong magnetic fields in the lab.
d) The visible wavelength is transparent to the cornea and lens.

CHAPTER 6 – DISCUSSION OF SOME PROCESSING TECHNIQUES

Q1) When using fiducials, it is always best to:

a) Use a high contrast fiducial mark.
b) Keep consistency and always use the SAME fiducial on any individual board.
c) Use a large fiducial so that the camera can find it.
d) Bury the fiducials under a cover layer to protect them.

Q2) Taper is sometimes a good thing to have. Which of the following illustrate positive aspects of taper in laser materials processing?

a) Taper is sometimes helpful in flow applications as it helps direct the flow medium (gas or liquid) through the orifice.
b) One can take advantage of the natural taper to drill holes smaller than what one would think possible using only simple calculations.

c) Many different tapered 'shapes' are possible to attain.
d) None of the above.

Q3) Assist gases perform all of the following EXCEPT?

a) Cool the plasma.
b) Help eject debris and vaporized material from the hole.
c) Keep oxygen away from the process area.
d) Stabilize the focus position with respect to the final focusing lens.

Q4) Which of the following are NOT important considerations when using assist gases?

a) The pressure of the gas.
b) The type of gas.
c) Coaxial delivery centered on the laser beam.
d) The size of the gas bottle.

Q5) A 50 micron high character is required on target. This character is to be made with a Gaussian beam laser delivered through a galvanometer beam delivery system. What is the maximum spot size on target needed in order to achieve this result?

a) 5 microns
b) 7 microns
c) 14 microns
d) 50 microns

Q6) Which laser in general cannot be used as a marking laser?

a) Excimer
b) CO_2
c) Fiber
d) DPSS

Q7) One of the problems encountered in scribing very thin films on plastic substrates is that the laser removes some substrate material. What method below could be used to minimize or eliminate this problem?

a) Increase pulse energy to increase peak power.
b) Use flat top optics to change the shape of the Gaussian beam.
c) Slow the galvo speed to increase dwell time.
d) Use a telescope to expand the beam.

CHAPTER 7 - APPLICATIONS

Q1) A printed circuit board has 1 oz. Cu ground layer, with 2 mils thick FR4 between it and the top layer of 1 oz. Cu. Microvias are required of diameter 200 microns drilled through the top layer of Cu, through the dielectric and stopping at the bottom Cu layer. Which method below would not be suitable to do this job?

a) Chemical etch top layer of Cu and use CO_2 laser to remove dielectric – a subsequent cleaning step is necessary.
b) Use a UV laser to drill through both layers and stop at the Cu – a subsequent cleaning stop is not necessary.
c) Use a UV laser to drill the top layer of Cu and a CO_2 laser to drill the dielectric. A subsequent cleaning step is necessary.
d) Use a CO_2 laser to drill the top layer of Cu and a UV laser to drill the dielectric – a subsequent cleaning step is not necessary.

Q2) Which laser would not be appropriate to scribe 25 micron wide lines in ITO coated on glass?

a) CO2
b) 532 nm
c) 355 nm
d) 266 nm

Q3) Referring to Figure 7.11, a 248 nm excimer laser is used to strip wires in a set up similar to that which is shown. The excimer laser puts out 300 mJ per pulse and the beam is 1 cm × 2 cm. The pulse length is 20 ns. The wire diameter is 100 microns and the coating is 10 micron thick polyimide. The strip length is 1 mm. If the energy density needed on target to ablate the polyimide without damaging the wire is 2 J/cm^2, what magnification telescope should be used to compress the beam?

a) 5.25
b) 3.65
c) 2.45
d) 2

Q4) In the example above, what if the strip length was 1 cm?

a) 5.25
b) 3.65
c) 13.3
d) 15.7

Q5) Diabetes test strips use several layers of material laminated together. All of the layers can in principle be laser processed, but for some of the layers – once a design is established – hard tools are made that are less costly in production. Estimate how long it would take to cut a 5 mm × 20 mm rectangular outline out of a plastic material assuming the spot size on target is 25 microns and approximately 0.1 microns of material are removed per pulse. The laser can be run at 100 kHz repetition rate. The plastic is 200 microns thick.

a) 15 seconds
b) 3 minutes
c) 150 minutes
d) 115 seconds

Q6) Using AOD's and end point detection, a matrix array of 500 holes can be drilled in thin polyimide in about 4 seconds. If the sensitivity of the detector decreases, what will happen?

a) The holes will get bigger.
b) The process will take longer.
c) Both of the above.
d) None of the above.

Q7) Microfluidics applications require precise and very small channels for fluid flow. These channels can be processed inside bulk materials if the right laser is used. What laser configuration below might work for this application?

a) CO_2 laser with galvo beam delivery
b) USP laser with partially transmissive material
c) 355 nm laser with highly absorbing material
d) Fiber laser with fixed beam delivery

Q8) A certain drug needs to be delivered to a localized area through a catheter. In order to deliver a certain volume of drug within a specified time to a specific location, a hole of diameter 75 microns was drilled. Unfortunately, the fluid pressure on the artery walls was found to be too much. What method below might be used to get around this problem?

a) Use a different drug.
b) Drill a matrix of smaller holes.
c) Increase the hole diameter.
d) Decrease the hole diameter.

Q9) Which of the below statements is false concerning the laser processing of non-homogeneous resin coated fiber materials like FR4 or Carbon epoxy fiber material?

a) Care must be taken to avoid heat transfer through the fibers.
b) Higher peak power is needed as the fibers normally require more energy density than the epoxy.
c) The process is faster than with homogeneous materials.
d) Undercutting may be a problem if the laser power is too high.

Q10) A laser has a pointing stability of +/- 5 mRad. A solar application calls for a 10 micron wide scribed line in a thin coating. This scribed line must be made at speeds of up to several meters per second. Can this laser be used effectively using an optical system with a 3 m pathlength?

a) Yes
b) No
c) There is not enough information to answer the question.
d) Lasers cannot be used for this delicate operation.

Q11) Which statement is true concerning LED liftoff with an excimer laser?

a) The substrate must be transparent.
b) The process requires multiple pulses per location.
c) Because of the LED structure, beam homogeneity is not important.
d) Each die is processed individually.

Q12) A surface needs to be textured with 10 micron 'dots' distributed in a random pattern over a surface area of about 2 square inches. Which laser would be the most appropriate for this job?

a) CO_2
b) 248 nm excimer
c) 355 nm DPSS
d) Fundamental fiber laser

Q13) USP lasers are being used in CIGS applications to remove the P1 Mo layer. Which of the below statements is true?

a) Collateral damage to the surrounding material is minimized with a short pulse laser.
b) Removal of thicker material is quicker and therefore imparts less heat into the surrounding material.
c) The lower peak power of this laser prevents damage to the substrate.
d) High repetition rate lasers are scanned slowly over the substrate in order to get the best edge quality.

Q14) Edge deletion using lasers is very typical in the solar industry when manufacturing thin film devices. Which statement below is NOT true of the edge deletion process?

a) It is done in order to perform the subsequent step of sealing against humidity.
b) It is dirty, even when using lasers in place of for instance sand blasting.
c) UV lasers are used in order to get the best cleanliness.
d) Pulse to pulse stability is crucial.

CHAPTER 8 – MATERIALS

Q1) Which of the materials below can be processed using a CO_2 laser?

a) Copper
b) Brass
c) Molybdenum
d) Teflon

Q2) The 1 micron wavelength fiber laser is particularly useful for which of the following?

a) Copper
b) Brass
c) Molybdenum
d) Teflon

Q3) When laser processing medical parts, an assist gas is frequently used. Which of the configurations below would benefit MOST from the use of assist gas?

a) Galvo processing using a 1 micron fiber laser.
b) Fixed beam processing using a 1 micron fiber laser.
c) Galvo processing using a 355 nm DPSS laser.
d) Fixed beam processing using a 355 nm DPSS laser.

Q4) A white catheter containing titanium dioxide pigment is marked with graduation lines around the circumference every 10 mm. In addition, alpha numeric characters are placed along one side indicating the value of the graduation. If the catheter is 3′ long and the marked area is about 24″, which laser would best be used for this application?

a) CO_2 laser
b) Fiber laser
c) 355 nm DPSS laser
d) USP laser

Q5) In the above example, what would be the best method for making the circumference graduations?

a) Mark along one side, then index the part (rotary) and mark again. Repeat until the shaft has been turned by 360 degrees, then mark the alpha numerics.
b) Rotate the part freely and jump the laser beam using a galvo to the proper point, then let the mark be made by free rotation of the shaft. When finished with all graduation marks, stop and mark the alpha numerics.
c) This is a two laser job – use the 355 nm DPSS laser for marking the graduations and the CO_2 laser for the alpha numerics.
d) TiO_2 does not mark well with any laser wavelength.

Q6) Again referring to the example given in #4 above, which of the below is NOT an assumption that has to be made?

a) A three axis galvo is used so that the entire 24″ marking field can be made without indexing along the catheter.
b) There is a defined end point so that the circumference graduations are all indexed from one end.
c) The energy density of the laser beam is below the ablation threshold of the material so no material removal occurs.
d) The part is place into position by an operator.

Q7) Teflon™ is a material that is sometimes difficult to machine with UV lasers. The processing can be enhanced if adulterants are added. What property should these adulterants have?

a) Low melting temperature so that the material will ablate cleanly.
b) High absorption at the wavelength being used.
c) High thermal conductivity.
d) Low moisture.

CHAPTER 9 – METROLOGY AND CLEANING

Q1) Which of the following does NOT fit with the rest?

a) 32.0×10^3 m/s
b) 1.05×10^4 ft/s

c) 1.26×10^6 in/sec
d) 5.78×10^7 ft/hr

Q2) A physical scrub done on laser processed parts has the following virtue:

a) It is easy and relatively inexpensive.
b) Care must be taken to avoid damage to the part.
c) Parts can be stacked and processed simultaneously.
d) Because it is a batch process, large parts require the same effort as small parts.

Q3) Electropolishing preferentially removes material in 'high spots'. What materials from the following list would be good candidates for electropolishing after laser processing?

a) Cleaning carbon residue from polyimide.
b) Removing 'burrs' from laser drilled copper.
c) Smoothing the rough surfaces after laser cutting glass.
d) Removing debris from laser processed silicone.

Index

D

T

U

V

W

Y

Z